Building a Sustainable Lean Culture

Continuous Improvement Series

Series Editors: Elizabeth A. Cudney and Tina Kanti Agustiady

Additive Manufacturing Change Management
Best Practices
David M. Dietrich, Michael Kenworthy, Elizabeth A. Cudney

Healthcare Affordability
Motivate People, Improve Processes, and Increase Performance
Paul Walter Odomirok, Sr.

Robust Quality
Powerful Integration of Data Science and Process Engineering
Rajesh Jugulum

Statistical Process Control
A Pragmatic Approach
Stephen Mundwiller

Transforming Organizations
One Process at a Time
Kathryn A. LeRoy

Continuous Improvement, Probability, and Statistics
Using Creative Hands-On Techniques
William Hooper

Affordability
Integrating Value, Customer, and Cost for Continuous Improvement
Paul Walter Odomirok, Sr.

Design for Six Sigma
A Practical Approach through Innovation
Elizabeth A. Cudney and Tina Kanti Agustiady

Building a Sustainable Lean Culture
An Implementation Guide
Tina Kanti Agustiady and Elizabeth A. Cudney

For more information about this series, please visit: https://www.crcpress.com/Continuous-Improvement-Series/book-series/CONIMPSER

Building a Sustainable Lean Culture

An Implementation Guide

Tina Kanti Agustiady
Elizabeth A. Cudney

CRC Press
Taylor & Francis Group
Boca Raton London New York

CRC Press is an imprint of the
Taylor & Francis Group, an **informa** business

First edition published 2023
by CRC Press
6000 Broken Sound Parkway NW, Suite 300, Boca Raton, FL 33487-2742

and by CRC Press
4 Park Square, Milton Park, Abingdon, Oxon, OX14 4RN

CRC Press is an imprint of Taylor & Francis Group, LLC

ISBN: 9781498798402 (hbk)
ISBN: 9781032287713 pbk)
ISBN: 9780429184123 (ebk)

DOI: 10.1201/9780429184123

Typeset in Palatino
by Deanta Global Publishing Services, Chennai, India

To my first-born child, Arie Agustiady. Your love for books makes me want to continue writing every step of the way!

To my princess, Meela Agustiady. You encourage me to be a better woman, mother, and professional!

To my dear husband, Andry, your love and support keep me motivated and driven!

Tina Agustiady

This book is dedicated to my husband, Brian, whose love and support keep me grounded and motivated.

To my handsome, thoughtful, and funny son, Joshua.

To my beautiful, talented, and driven daughter, Caroline.

I love you with all my heart!

Beth Cudney

Contents

Preface

This book is intended to be an implementation guide for creating a Lean culture from the ground up while gaining buy-in from key stakeholders and sustaining the results. Lean is widely used in a multitude of industries such as manufacturing, service, healthcare, IT, utilities, government, and aerospace, among many others. Lean training often talks about implementing a Lean culture; however, typically only Lean tools are discussed for the implementation. This book discusses how to implement the entire Lean system from three main aspects:

1. Management support
2. Hoshin planning
3. Continuous improvement tools

Lean is a continuous improvement philosophy that must be embraced as an enterprise solution. The Lean journey involves a cultural change throughout the organization to drive continuous improvement efforts. In order for Lean implementation to be successful, management must fully support and be engaged throughout Lean implementation. Lean leadership is necessary to ensure the Lean journey does not fail through accountability, standardized work, and structured meetings with outcomes. This involves active coaching and mentoring.

Further, Lean efforts must be integrated with the long-term strategic goals of the organization in order to gain buy-in and support. Hoshin Planning, also known as Hoshin Kanri or Policy Deployment, incorporates how management designs their goals at the top level and how they flow down. The implementation of Hoshin Planning ensures the strategic goals of a company drive progress and action at every level while reducing waste from inconsistencies due to poor communication.

While Lean is a philosophy, Lean also encompasses several tools that help eliminate wastes and standardize processes. Therefore, problem-solving methodologies are employed to identify and eliminate wastes. In addition, visual management is used to ensure clear and consistent communication.

Lean efforts should focus on empowering employees to want to build a Lean culture.

Acknowledgments

Our thanks and appreciation go to all of the Lean and Six Sigma team members, project champions, and mentors that worked so diligently and courageously on continuous improvement projects.

We would also like to give a special thank you to several people for their contributions to the development and production of the book at CRC Press/Taylor & Francis Group, including Cindy Renee Carelli (Senior Editor), Jennifer Ahringer (Editorial Assistant), and Jayanthi Chander (Project Manager).

Authors

Tina Kanti Agustiady is a certified Six Sigma Master Black Belt and Continuous Improvement Leader. She is currently a vice president of Lean Training and Development at JP Morgan Chase.

Agustiady worked previously at MetLife across various deployments as part of leading sustainability efforts for MetLife Way. Agustiady oversaw sustaining Lean transformations and creating a Kaizen Practitioner Certification Program.

Agustiady was responsible for the strategic and tactical implementation of a Lean system deployment and furthering the transformation to a Lean culture to support strategic business initiatives and help drive overall operational excellence within the continuous improvement initiative for a US$ 2-billion manufacturer of construction components by focusing on management of cross-functional teams and improvement of operational metrics and objectives at all levels of the organization.

As director, operations Master Black Belt at Philips Healthcare, Agustiady drove all continuous improvement projects in the CT/AMI Operations function, resulting in the highest efficiency and effectiveness levels within Philips Healthcare. She was the transformation leader for the two businesses providing coaching and leadership in the new methodology. She was recently a strategic change agent as a key member of the BASF Site Leadership Team responsible for infusing the use of Lean Six Sigma throughout the organization.

Agustiady consistently improves cost, quality, and delivery by applying Lean and Six Sigma tools to achieve improvements through a simplification process. She is an experienced leader who has facilitated many Kaizen, 5s, and Root Cause Analysis events throughout her career in the healthcare, food, and chemical industries. She has conducted Six Sigma training and improvement programs in the baking industry at Dawn Foods and at Nestlé Prepared Foods where she held positions as a Six Sigma product and process design specialist responsible for driving optimum fit of product design and current manufacturing process capability, reducing total manufacturing cost and consumer complaints.

Her many activities and responsibilities include serving as the past Lean Division president and current technical vice president of the Institute of Industrial and Systems Engineers (IISE). She has served as a board director and chairman for the IISE annual conferences and Lean Six Sigma conferences.

Agustiady is an instructor who trains and certifies students for Lean Six Sigma for IISE.

Her accomplishments as a writer and author are numerous, including serving as an editor for the *International Journal of Six Sigma and Competitive Advantage*. She has co-authored *Statistical Techniques for Project Control, Sustainability: Utilizing Lean Six Sigma Techniques*, and *Total Productive Maintenance: Strategies and Implementation Guide*." She has also authored *Communication for Continuous Improvement Projects* and her recently published book, *Design for Six Sigma: A Practical Approach through Innovation*. She serves as series editor for the CRC Press/Taylor & Francis Group book series Continuous Improvement.

Agustiady was a featured author in 2014: http://www.crcpress.com/authors/i7078-tina-agustiady.

She was honored to be a Feigenbaum Medal winner for 2016, presented to an individual who is 35 years of age or younger (as of October 1 of applying year), who has displayed outstanding characteristics of leadership, professionalism, and potential in the field of quality and also whose work has been or will become of distinct benefit to mankind: http://asq.org/about-asq/awards/honors/feigenbaum.html.

Last but not least, Agustiady is the winner of the 2018 ASQ Crosby Medal, presented to the individual who has authored a distinguished book contributing significantly to the extension of the philosophy and application of the principles, methods, or techniques of quality management: https://asq.org/about-asq/asq-awards/honors/crosby.

Agustiady received her BS in industrial and manufacturing systems engineering from Ohio University. She earned her Black Belt and Master Black Belt certifications at Clemson University.

Elizabeth (Beth) Cudney, PhD, is a professor of Data Analytics in the John E. Simon School of Business at Maryville University. Dr. Cudney received her BS in industrial engineering from North Carolina State University. She received her Master of Engineering in mechanical engineering with a manufacturing specialization and Master of Business Administration from the University of Hartford, and her doctorate in engineering management from the University of Missouri – Rolla. Her doctoral research focused on pattern recognition and developed a methodology for prediction in multivariate analysis. Dr. Cudney's research was recognized with the 2007 ASEM Outstanding Dissertation Award.

Dr. Cudney has published 9 books and over 200 refereed publications. She is an ASQ Certified Six Sigma Black Belt, Certified Quality Engineer, Manager of Quality/Operational Excellence, Certified Quality Inspector, Certified Quality Improvement Associate, Certified Quality Technician, and Certified Quality Process Analyst. She is also an IISE Certified Lean Six Sigma Master Black Belt.

Dr. Cudney received the 2022 ASQ Crosby Medal for her book on Lean Six Sigma. Dr. Cudney received the 2021 Bernard R. Sarchet Award from the Engineering Management Division of the American Society for Engineering Education for "lifetime achievement in engineering management education." She was also awarded the 2021 Institute of Industrial and Systems Engineers Operational Excellence Division Teaching Award. She received the 2020 Walter E. Masing Book Prize from the International Academy for Quality for her book on Lean Six Sigma. In 2018, Dr. Cudney received the ASQ Crosby Medal for her book on Design for Six Sigma. Dr. Cudney received the 2018 IISE Fellow Award. She also received the 2017 Yoshio Kondo Academic Research Prize from the International Academy for Quality for sustained performance in exceptional published works. In 2014, Dr. Cudney was elected as an ASEM Fellow. In 2013, Dr. Cudney was elected as an ASQ Fellow. In 2010, Dr. Cudney was inducted into the International Academy for Quality. She received the 2008 ASQ A.V. Feigenbaum Medal. This international award is given annually to one individual "who has displayed outstanding characteristics of leadership, professionalism, and potential in the field of quality and also whose work has been or, will become of distinct benefit to mankind." In addition, she received the 2006 Society of Manufacturing Engineers (SME) Outstanding Young Manufacturing Engineer Award. This international award is given annually to engineers "who have made exceptional contributions and accomplishments in the manufacturing industry."

1

Introduction to Lean and the Importance of Cultural Change

A strategy that is at odds with a company's culture is doomed. Culture trumps strategy every time.

—Jon R. Katzenbach, Ilona Steffen, and Caroline Kronley

Lean Overview

"Lean" is a terminology that is well known across industries. Lean is defined as the elimination of waste in operations and processes through managerial principles. Many principles are comprised in the Lean concept, but the main concept is the effective utilization of resources and time in order to achieve higher quality products and ensure customer satisfaction.

It is important to remember that Lean is a choice. Nobody can make Lean happen. It is what you make of it. You can stay stagnant and keep doing things the way they have always been done. Or you can seek to continuously improve. The most important part of Lean is THE PEOPLE. Lean is not just about reducing waste and improving flow, but empowering the people by listening to the people. Understanding the people. Without the people, the business is not a business.

Continuous improvement refers to all aspects of improving processes, products, methodologies, and techniques to ensure they provide more value to customers and are sustainable.

Continuous improvement is about working smarter and not harder. This is the definition of Lean: The pursuit of **perfection** using a systematic approach to **identifying** and **eliminating waste** through **continuous improvement** of the value stream, which enables the product or information to **flow** at a rate determined by the **pull** of the **customer**.

We are trying to eliminate waste by continually improving using Lean tools to satisfy the customer.

DOI: 10.1201/9780429184123-1

Some may have bad misconceptions of Lean:

- Lean is NOT the next headcount reduction exercise.
- BUT: Lean creates **opportunities** for doing more value-added activities.
- Lean will not succeed if the initiative stays limited to "operations."
- Lean is NOT about working harder, but rather working smarter.

Traditional companies have some effectiveness that goes up and down over time. It is irregular, and there is no standardization or sustainability. Lean organizations strive to be the best by continually improving. Lean organizations have standardization, so there is sustainability and continuous improvement. This is the satisfaction model for the customer. As quality goes up so does delivery while costs go down. We must provide quality on time while reducing costs in order for the customer to be happy.

According to Toyota:

Quality is inherent in Toyota's products.

The company is constantly striving for improvement (Kaizen), which has direct benefits for their customers. Toyota's insistence on maintaining quality throughout the production process is vital to ensuring that their finished products are of the highest quality.

Cost is always an issue.

By buying Toyota products, their customers can be sure they have made a good choice. Kaizen ensures that Toyota products feature the latest effective innovations, while maximizing productivity. The quality of Toyota's products allows their customers to enjoy a high return on their investment.

Delivery is right for each customer's order.

Toyota's customer-driven system ensures that production output corresponds with timely delivery. Toyota's smooth, continuous, and optimized workflows, with carefully planned and measured work-cycle times and on-demand movement of goods, allow them to consistently meet their customer's expectations.

These three areas are illustrated in the Time, Cost, Scope, Quality Triangle (Figure 1.1).

There are a few key concepts of Lean.

First, it is critical that we understand the customer and what exactly it is that they want. Then we focus on creating a continuous flow of production where we are not interrupting our processes with waste. Next, we must pull materials from upstream. Each process pulls for the next.

We must always eliminate waste. Waste is any non-value-added activity.

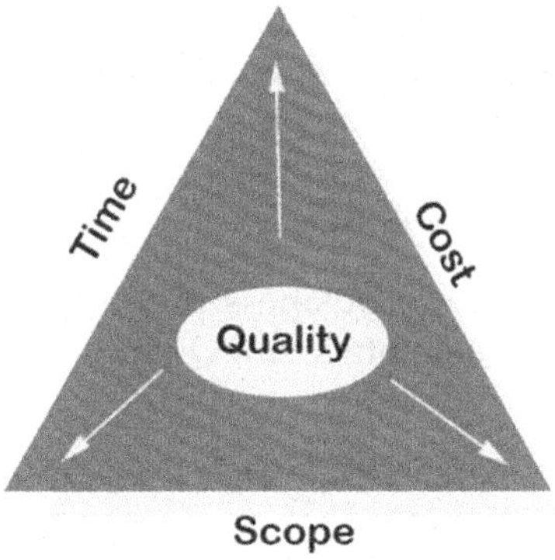

FIGURE 1.1
Time, Cost, Scope, Quality Triangle.

A typical house of Lean has the following:

It starts with the foundation – 5S, cleanliness, Kaizen to continuously improve, total productive maintenance – to ensure we have quality maintenance processes.

Then there are the pillars of our house.

Just-in-Time (JIT) is as follows:

Pull – There are three basic types of pull system: replenishment pull, sequential pull, and mixed pull system with elements of the previous two combined (see glossary at the end of chapter). In all three cases, the important technical elements for systems to succeed are: (1) flowing product in small batches (approaching one piece flow where possible); (2) pacing the processes to takt time (to stop overproduction); (3) signaling replenishment via a kanban signal; (4) leveling of product mix and quantity over time.

Flow – A continuous flow process is a method of manufacturing that aims to move a single unit in each step of a process, rather than treating units as batches for each step.

Takt time – Takt time is the rate at which work must be performed for customer demand to be met on time.

Heijunka is leveling, and this is done in order to meet demand while reducing waste.

A Lean cell design ensures waste is minimized and the process flows using a pull system.

Single minute exchange of dies (SMED) will minimize changeovers.

The middle pillar is the most important because it is all about the people. We need to build teams, empower people through cross training. We must understand the management vision through Hoshin Planning and finally understand the supplier and have a good relationship with them. Hoshin

Kanri is a planning and implementation process which gives direction to an organization when looking at future strategies. "Ho" means direction, "Shin" means needle. "Hoshin" means compass. "Kan" means control or channeling. "Ri" means reason or logic.

Hoshin Planning does the following:

- Facilitates the creation of business processes that result in sustained competitive advantage in Quality, Delivery, Cost, and Innovation
- Aligns the major strategy objectives with specific resources and action plans
- Consists of a five-step process beginning with high-level strategic objectives and ends with local-level improvement targets
- Utilizes term called "Catchball" which means driving force of alignment, clarification, and employee involvement

Finally, there is Jidoka which is quality at the source. Poka-yokes eliminate defects from happening in the first place. Andons are signals to signal a defect or problem. Autonomation empowers employees to build in quality every day. We must ask five why's to understand the root cause. We must stop the line when we have a problem so we can fix it right away. Built in quality will help us finalize the pillar.

All of this leads to a happy customer and a successful business.

The House of Lean is illustrated in Figure 1.2.

Lean systematically aligns people and processes with our strategy.

- Lean is the elimination of waste to improve the flow of information and material.
- What happens when you don't eliminate waste?
  - It adds cost to the product/service with no corresponding benefit.
  - It destroys competitive advantage.
  - It makes work frustrating.
  - It uses valuable resources (i.e., your time) to produce no value.

Value-added time is anything the customer is willing to pay for.

Non-value-added time is anything that does not add form, feature, or function which the customer does not want to pay for.

There are three main types of waste: Mura, Muri, and Muda.

- Mura is unevenness.
- Muri is overburden.
- Muda is pure waste.

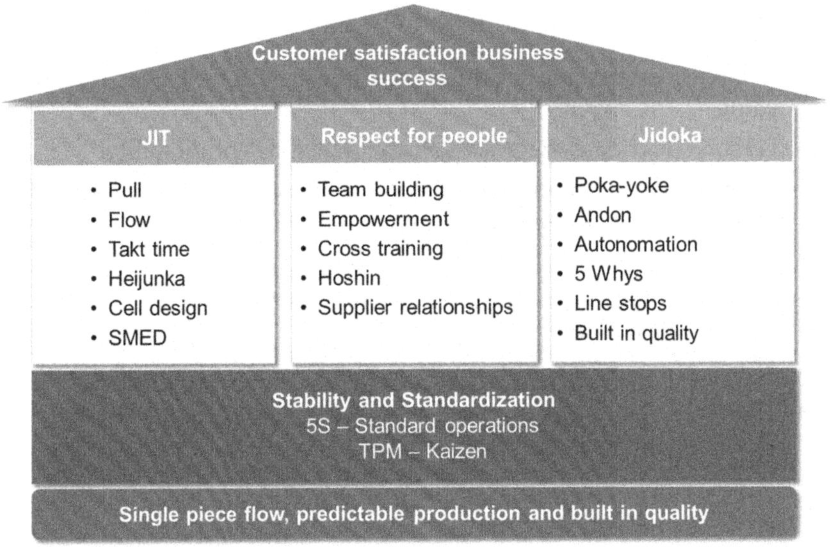

FIGURE 1.2
House of Lean.

In Figure 1.3, waste shows the three main types of waste:

Mura is considered unevenness, while Muri is overburden and Muda is non-value-added activities.

Mura can be solved with proper forecasting techniques.

Muri can be solved with proper line balancing techniques.

Waste should be looked for through all processes and ultimately removed.

FIGURE 1.3
Waste.

Muda is:

- A Japanese term for anything that is wasteful and doesn't add value.
- Waste reduction is an effective way to increase profitability.
- Waste occurs when more resources are consumed than are necessary to produce the goods or provide the service.
- Anything that doesn't add value to the process.
- Anything that doesn't help create conformance to your customer's specifications.
- Anything your customer would be unwilling to pay you to do.
- There are eight main types of wastes. We use the acronym DOWNTIME to explain the wastes:
    - Defects
    - Overproduction
    - Waiting
    - Non-utilized talent
    - Transportation
    - Inventory
    - Motion
    - Excessive processing
- Here are examples of the eight types of wastes:
    - **Defects** – Efforts caused by rework, scrap, and incorrect information
    - **Overproduction** – Production that is more than needed
    - **Waiting** – Wasted time waiting for the next step in a process
    - **Non-utilized Talent** – Underutilizing people's talents, skills, and knowledge
    - **Transportation** – Unnecessary movements of products and materials
    - **Inventory** – Excess products and materials not being processed
    - **Motion** – Unnecessary movements by people (e.g., walking)
    - **Excessive processing** – More work or higher quality than is required by the customer

Let's discuss a few examples of the eight wastes in manufacturing, supply management, and service. I'm sure you can relate to many of these wastes.

WASTES	MANUFACTURING	SUPPLY MGMT	SERVICE
Defects	Rework, scrap, poor quality	Missing/wrong supplies	Errors, misinformation
Overproduction	Unclear excess production	Excessive warehousing	Information overload
Waiting	Waiting, delays, idle time	Order/delivery delays	Delays, meeting overruns
Non-utilized talent	Unused resources/skills	Under-utilizing capabilities	Wrong resource allocation
Transportation	Transport of materials	Small quantity deliveries	Travel/search activities
Inventory	Work in progress, parts	Overstocked supplies	Excessive multitasking
Motion	Poor production layout	Difficult government approvals	Unnecessary action
Extra processing	Overshooting requirements	Excessive documentation	Duplication/excess work; approvals

FIGURE 1.4
Waste explained.

Think about even creating multiple Excel spreadsheets for different people that ultimately provide the same information. Could we consolidate? See Figure 1.4. for examples.

No two Lean implementation plans are ever the same. They require support processes and strong tools. Piloting initiatives show where mistakes are prevalent first hand and where improvements can be made.

Importance of Cultural Change

Change involves yourself changing along with others. Change can always be for the better, but no change at all can never lead to better situations. Change management relies on the understanding of why things are done and why people are comfortable. Managing others through change processes is needed to change the status quo. This type of change needs guidance, encouragement, empowerment, and support.

What is the status quo anyway? The status quo is defined as the existing state of affairs. In Latin, the meaning is "the state in which." Therefore, maintaining the status quo means to keep things the way they currently are. Some people have the mentality of "If it isn't broken, why fix it?" This methodology can never lead to change or success because even if things are going well at the present time, in due time other changes in the world will come into effect that make things not go as well as planned. In a business setting, every corporation strives to be the best it can be in what it does. The competitors then try to beat the best-in-class corporation by doing things differently and better than their competitor. Eventually, the best-in-class corporation is the one that produces the most satisfying changes, but the change must be present in order to satisfy their customers. Keeping things the same way rarely satisfies customers because customers can become complacent and their needs change. Humans desire change and innovation. The desire to be different motivates others to change the status quo. Being complacent normally means being safe and avoiding controversy. Even though this is a safe measure, it will not end with a best-in-class way of doing things because complacency becomes tiresome and the thrill of excitement is taken away.

Within Lean, the culture of the organization must change for the techniques and tools to be effective. The status quo changes from a mentality of fixing things when they are broken to being more proactive. In addition, Lean involves everyone within the organization. Therefore, the organization changes from a silo approach based on function to everyone taking an active role in improving products and processes. This change can cause fear in employees. Employees may fear change because they may feel they are losing ownership. While others may fear the unknown, since they are being asked to take on new tasks and responsibilities outside of their comfort zone.

Effectively implementing change involves frequently incorporating new competencies. New competencies will enable further education and training for all employees. Once this new mentality becomes the norm, changes are viewed as good and they are always anticipated. The changes will begin from simple items to making work easier to strategic thinking where daunting tasks are eliminated by means of more technologically advanced methodologies.

People are afraid of change because it makes comfortable ways uncomfortable while changing the normal way of operating. This needs to be considered when implementing Lean. Before Lean is rolled out, these fears should be addressed in developing a communication plan with the Lean rollout. What should be taught about change is the fact that new skill sets are being taught, which opens doors and increases continuous education. The first reaction of asking someone to change can be taken negatively because they are being told to stop doing what they have always been doing. Recardo (1995) developed a comprehensive list of why change fails:

- The organization's architecture is not aligned and integrated with a customer-focused business strategy.
- The individual/group is negatively affected.
- Expectations are not clearly communicated.
- Employees perceive more work with fewer opportunities.
- Change requires altering a long-standing habit.
- Relationships harbor unresolved past resentments.

- Employees fear a loss of job security.
- No accountability or ownership of results.
- Poor internal communications.
- Insufficient resource allocation.
- It was poorly introduced.
- Inadequate monitoring and assessment of change progress.
- Disempowerment.
- Sabotage.

These are all aspects of fear that should be considered with Lean. To help people understand why the change is needed, the approach to the Lean methodology should be given along with the reason the change is desired. Most of the time change is needed so that the organization can operate more efficiently while speeding up tasks that need to be completed. Understanding the organization's need for change is imperative in order to embrace the change. This helps the individuals to also understand "what's in it for me?" (WIIFM).

Heller (1999) developed a model for managing change, which is shown in Figure 1.5. The response to change is based on how individuals handle the change. Individuals will go through each type of response, although some individuals may take longer in the different response stages. By developing a communication plan and effectively communicating the need for change and how it will affect individuals, people can move quicker to acceptance.

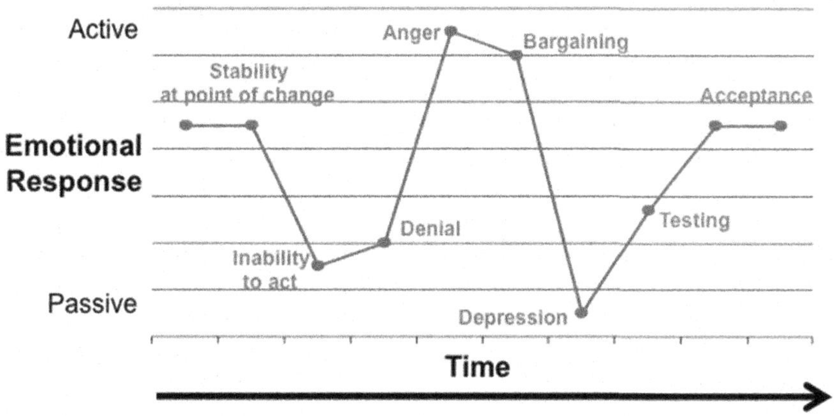

FIGURE 1.5
Change management model.

In addition, there are several reasons why people resist change including:

- Desire not to lose something of value
- Misunderstanding of the change and its implications
- A belief that the change does not make sense
- Low tolerance for change
- Fear especially of the unknown that change represents
- Habit and inertia – it is easier to stay the same
- No ownership of the problem
- Not a priority
- Do not understand it
- Lack of skills needed for the change
- Loss of power

Change requires involved leadership with strong communication skills. The following must be provided while implementing Lean:

- A clear direction and purpose of what is happening along with focused goals and rationale for the changes being made.
- People must be willing to let others make mistakes and learn from them without being fearful of blame.
- Being able to express their feelings openly encourages communication and honors credit for good ideas, good work, and successes.

As you implement Lean in your organization, you will undoubtedly have individuals that have heard about Lean, and based on that little information or gossip, they may have heard very positive or very negative things. Spotting resistance to change is important because negative connotations to change must be eliminated. Spotting change comes from listening to gossip or what people have to say about why things will not work. Observations of people will tell whether the employees are resisting change as well. If employees are being extremely negative, arguing with others no matter what statement is made, missing deadlines or meetings, or not attending or reading up on change assignments, it normally means they are resistant to change. This resistance must be minimized by addressing the situation quickly. The resistance can only be minimized if the employees have trust in the changes. If the employees are involved in the decision-making and receive consistent and frequent communication, they become less resistant to the change because they were decision-makers in the changes made which creates buy-in. Direct, frequent, and consistent communication also gives visibility on the changes happening with less ambiguity on why the decisions were made to implement Lean. Building positive relationships helps minimize this change by having positive bonds of people working together. The multiple positive attitudes end up being passed on to others with negative attitudes making the entire workforce change. The open exchange of ideas must be able to occur in order to change. If employees are afraid to bring up their ideas on Lean, they will continue to resist change because they are not being heard. It is essential to gain employees' trust and allow them to speak their minds. Once they describe the feelings they are having and the reasoning behind them, it is important to understand the position they are in. By minimizing the fears your employees have from Lean implementation, you will slowly gain their trust and ensure their fears are not going to come true.

In addition, there are several required elements for change:

- Change objectives
- Change vision
- Champion, sponsors, facilitators, teams
- Training
- Systems that enable and support Lean
- Lean performance metric(s)

Knowing what is driving the change, the need for Lean, and how it will affect individuals and the organization is an important step to acceptance of change. Once the change has been adapted, people will begin to accept the change and be willing to make more changes for the better.

Heller (1999) developed several strategies for overcoming resistance including:

- Expect resistance
- Create allies
- Make the case for change
- Choose your opening moves carefully and make them bold
- Articulate the destination; over communicate
- Resolve "me" issues
- Involve people to increase commitment
- Utilize resources effectively
- Promise problems
- Over communicate
- Beware of bureaucracy
- Get resistance out in the open

- Give people the know-how needed
- Role model and embody the change values
- Measure results
- Reward change
- Outrun the resisters

Being a role model during change eases the difficulty of the change. People want to be facilitated through change transitions and need guidance and coaching. To be a role model during the Lean implementation, it is important to put yourself into the other person's shoes that needs to complete the change. Thinking about all the reasons the change can be bad will help with the understanding of how others are feeling and coping. Giving the business vision of why the change needs to happen will also help guide the scope of the project. Never accept "That's the way it has always been done" as an answer. This answer is an excuse and not a solution. The vision must be clearly defined and provide a direction that everyone understands. It should focus on measurable goals, include the demand for the change, appeal to various learning styles, and answer the "me" questions.

The combination of skill sets will help with challenging the status quo. Different people have different ideas and combining the ideas to come to a uniform decision will help the group to find strategic opportunities. The group needs to be flexible during the Lean implementation, and one individual cannot overpower the conversation or ideas being given. Adaptability of other people's viewpoints, schedules, concerns, and styles should be taken into consideration. No one idea is right. Showing other people as a leader that you are open to change will help other people appreciate your values and anticipate the change. Limiting expectations for a perfect workday scenario should be considered in order to accommodate other people and give them the time and attention they need during the time of change. People need a great deal of reassurance during changing times, and as a leader that reassurance and commitment need to be present daily. Leading with values such that others will respect and follow you is important because people do not follow others just because they have the title of manager. The vision needs to be not just told, but shared and agreed on by all. Leaders need to show that they are not changing the status quo for selfish reasons but for the benefit of the project and the business.

There are several success factors with implementing Lean. First, it is important to prepare and motivate people. This involves a widespread orientation to Lean and the necessary training. Management must create a common understanding of the need to implement Lean. Second, there must be employee involvement. As you implement Lean, try to push decision-making and Lean system development down to the lowest levels. As employees are involved with developing Lean, this will also drive buy-in. Next, it is important to share information and manage expectations throughout the entire Lean implementation and once it is in use. Finally, it is beneficial to identify and empower champions for Lean.

Prior to the Lean implementation, all employees need to understand their new roles. Timelines and targets for the implementation should be established and communicated. It is also important to establish easy to achieve near-term goals to provide opportunities for success early on.

The timing of change is important as well to consider with Lean. People need time to understand the change, and let the new thoughts and ideas sink in. During this time, many questions will incur, and employees should be able to speak with their managers about the questions they have with an open mindset mentality. Employees need to be a priority to managers. Without employees, managers are not managers. Competing priorities will lead managers to have little time for their employees. Clarifying expectations of changes and what you as a manager are working on will help others understand what they are dealing with. Employees should never feel as though they cannot talk to their managers. Once this fear is instilled in their minds, the trust has already been broken and mending the trust is harder to fix than starting with an open mindset mentality. If priorities do change, the transparency should be given promptly with explanations for the changes in priorities. All people are emotional, so giving facts about situations rather than reacting toward emotions is important.

During the Lean implementation, you should anticipate and celebrate successful results to maintain positivity. This positive feeling can be contagious. The approachability of the changes can be taken from different aspects:

- Positive thoughts approach
- Humoristic approach
- Realistic approach
- Open-minded approach

Maintaining a positive attitude is the most important aspect of changing. Positivity will gain respect and will allow others to see the brighter side of the picture. Humor can ease tensions and bring people together by seeing the lighter side of a situation. Humor helps during the beginning of relationships being established to show there is a fun and real side of perspectives. Being realistic helps people not to gain false expectations of what could occur and honesty is well presented. Being open-minded helps people see all versions of what could happen and appreciate different points of view.

Resistance to the Status Quo

All approaches and reactions are different and should be seen with an open mind. Most people resist the status quo versus embracing it. This is no different with implementing Lean. This is a problem that should be overcome and should be seen with an end in sight. Ensuring people understand the benefits in the change helps to fight the resistance. Incorporating different views and ensuring all perspectives are taken into consideration will help decisions to be made and accepted. Motivating and encouraging others that changing the status quo for the better will help resistance as well. Listening and understanding through the resistance will also show that the different perspectives are appreciated versus having a "my way because I said so" attitude. Patience is extremely important through resistance. It may take people more than an hour or a week or longer to embrace change. Showing that you have patience through the change will help ease the situation and show that there is a realistic approach behind the methodology. It is important to show, share, and ensure others believe the vision being handed at large. Alignment is key to changing systems and resistance. Creating lists of pros and cons can also show the vision and reasoning behind changing. During the list creations, the current and new ideas should also be listed to show why things were performed in the past and what benefits Lean will provide for the organization and the employees. Finally, positive reinforcement is essential through resistance of change. Not all people have natural self-confidence. Giving people the confidence they need is easy and is a coaching practice that makes good leaders. Publicizing wins helps grow people's self-confidence and shows working through the difficulties will be managed appropriately. Flexibility through this management must be learned by employees, manager, and leaders.

Utilizing Known Leaders to Challenge the Status Quo

Self-leaders can be identified by their ability to influence others and engage in activities. It is to a manager's benefit to utilize these leaders to help the change. Assessing the current structure of whom the natural leaders are, whom the resistance comes from, and who the followers may be will make the process easier to understand and manage. Simple questions can be asked:

- Who do people listen to?
- Who will adapt to change easily?
- Who will resist change?
- Who will help with the alignment of change?
- Who will motivate others?

Once the support structure is understood, the utilization of the proper people in the proper aspects will align the process change management.

Identification of trouble areas will help leaders to be realistic in what items may not go as planned. Utilizing the known leaders to communicate and distribute information needed will also help alleviate problems. These people will assist in supporting the structure and processes and maintaining the vision. Goal alignment with known leaders must be established before the leaders are utilized to motivate others. Once these goals are shared and agreed upon, these leaders will help change management drastically.

Communicating Change

Communication is the most important aspect to successful changes.

Communication prepares others for what is to happen, creates shared and agreed upon visions, and builds relationships within groups. Communicating change should consist of many aspects. Communicating specifics of what changes are incurring is important for people to understand to gain their buy-in. People need to understand the importance of change to the business and the reasoning behind it. It is important for the person delivering that message to understand themselves the reason of the change. The downfalls of what could happen if the change does not occur should be explained at this point in time as well. Explaining the urgency of the change to occur will speed up progress. Specifics on timelines and milestones also help drive the urgency.

Motivating people to change is imperative so they feel empowered to change and do not just make changes that will not be able to be sustained.

Finally, publicize the wins of work well done or employees that have embraced change. Preparing information and updates on results, changes, and progress will help people visualize the impact the changes are making. It is important to communicate as many of these wins in person. Many people will begin asking questions; admitting not knowing information is acceptable as long as there are means to showing you will find out the information. Being truthful and transparent will help expectations more clear even when discussing the downfalls.

Re-capping of past communications and topics with outcomes will help gain credibility on the work that was performed along with the successes that can be seen.

Many people will want to understand their roles and responsibilities; showing favoritism should not occur. Consistency is a key for success when dealing with different people and different personalities. Mistakes will happen, so learning from mistakes while being transparent about why mistakes incurred will prevent them from happening in the future. Capturing best practices and learning techniques will help benchmark ways for working in the future.

Engagement from all levels of employees will help the support of projects and increase awareness of new initiatives.

Successful transformations and Lean implementations come from asking the basic questions to gain clarity through the 5 whys and 1 H (who, what, where, why, when, and how). How to change and what to change are the big factors that will explain the success of change. Once the establishments of these measures are in place, the implementation process is simplified. Successful workplace changes occur with the proper foundation and framework that has been established through management and employees through a shared vision. People must drive these visions together with their different skill sets. The cross-functional groups working together as teams are what drive successful change. The real change comes from the people combined together.

Information sharing during the Lean implementation is a powerful means for communication. Information given to one another builds teams to share known information and makes strategic decisions from the data. A company's culture is based on this information sharing. Culture is based on beliefs and values, and the combination and agreement of the two. Wanting to increase productivity and efficiency is part of the culture. If all people believe in this, the vision is the same. If some people have not bought in to the vision or need to change, then a culture shift needs to take place. Once the culture has the same enterprises of beliefs, the overall vision can be achieved by all entities working on the same goals due to their beliefs.

According to The Steve Roesler Group, Heller (1999) there are five clear-cut messages that show it is time for a change. These five qualities must be looked for and are defined next:

- *People whom you trust strongly believe you should make a change* – If multiple people that you are close to think it is time for things to be different, it may be a time to listen to them. They may understand you and see the changes may actually benefit you.
- *You're holding on to something and just can't let go* – If a particular situation is continually on your mind and you simply cannot let it go, this is seen as a signal. If this is bothersome for you, instead of mentally abusing yourself, a change could ease your mind.
- *You feel envious of what other people have achieved* – Jealousy is an evil beast, but it may be able to help you better yourself. If you are envious of someone or something, instead of being envious, you are able to change your ways to become more successful instead of simply being jealous. Taking action toward bettering yourself reduces jealousy and makes you proud of yourself.
- *You deny any problem and are angry in the process* – Anger is seen as a symptom of denial. Looking for help from someone or being able

to help someone in need reduces the anger. Increasing communication will help mitigate the problems by allowing individuals to admit there is a problem at hand. Having an open mind to the problems will encourage you to make a change for the better.

- *If you do absolutely nothing, the problem will continue* – Without addressing the situation or holding people accountable, no change will be made. This is because it is not known that the problem is bothering you, or nobody wants to address the problem at hand. Addressing the situation must happen for change to occur, but must be in a professional and calm manner, or will be seen as negative. Being honest and up front will help others see what changes can be made and will help you be able to change for the better with the communication received.

There are several simple methodologies for tackling workplace status quo.

Communicating where wasted time is being spent – This involves giving facts on costs and data behind efficiencies. Visual management of key performance indicators (KPIs) will address where targets and goals are.

Address costs from doing work repetitively – Discuss extra labor hours and extra materials spent due to mistakes and having to correct these mistakes.

Communicate the need for efficiency improvements – Companies and people generally need to be more efficient. Communicating the need to increase efficiency because capacity restraints are present is an easy way to show the goals of efficiency improvement are needed. Showing competitor's efficiency numbers or predictions will help communicate the baseline of where the efficiency should be.

- *Show that change is easier than people think* – Nobody likes to have a "flavor of the month" of how to do or address something. People must see change as structured and positive. It is important to show change is innovative and teaches others creative measures. Continuous education is imperative to growing, and changing the status quo will invigorate these behaviors by having an intelligent workforce that is up to date on technology and teachings.
- *Show that change is a habit* – Change should be a daily habit and not a way of doing something for a particular time. Incorporate change into people's everyday lives so that change becomes something easy and normal. Show changes are good for personal and professional growth, and must be incorporated in their work life and personal life.

Dealing with change is simple if people trust why changes are occurring. Explaining that change is needed in order to be competitive and expand while being innovative must occur for the change to take place. Management's beliefs and values need to be shared as a combined vision versus a top-down approach so that each individual is on the same page about change. There are four main changes that occur in the workplace:

- New products or services
- Organizational changes
- New management
- New technology

Explaining the need for the different changes and the rationale behind the decisions made will help buy-in for the change. Sometimes, the change does not seem to be for the better (e.g., layoffs). It is important when communicating change, even change that is not desired, that a positive outlook can be seen from the change. The best way of communicating any type of change is through factual information to the best information given. The majority of the time, the changes taken place is positive and affects cost, processes, and culture.

Cost changes should be discussed as being for the better in order to be competitive in the market place. If cost changes do not occur, the market demand will go toward the cheaper good especially if there is a good choice of quality. Process changes must be made to have continuous improvement. Increasing efficiencies will increase the market for being competitive as well while accomplishing an easier way of doing things. Innovation should be sought after to keep up on the latest technologies. Cultural changes are normally the most difficult because it changes the way normal operating procedures take place. Explain that cultural changes are needed, so there is a combined vision between all people in the workforce and there is no divide between management and other employees.

Implementing successful change involves utilizing the proper resources whether those resources are things or people. The plan then needs to be established and agreed upon so there is reasoning behind the changes being made. The changes then need to be implemented. Many times people think of great improvement ideas but do not fully implement them making the sustainment of them impossible. Finally, communication is the end of the change management. All steps through the process must be communicated to all people and must be agreed upon in order to be successful. Effective communication involves communicating often, giving reasons behind implementing Lean; explaining who was involved in the changes; and giving status updates of what changes have occurred, what will occur, and the effect on the business from the changes.

Conclusion

As a rule of thumb, plan for risks to occur and have a risk management plan. Risks can be threats or opportunities. Brainstorming potential risks is a key way to begin the problem-solving process. Risk actions can consist of avoiding risks, mitigating risks, acknowledging risks, or accepting risks. There are many tools and techniques that can help mitigate risks, which include decision trees, SWOT analysis (Strengths, Weaknesses, Opportunities, and Threats), risk grids, cause and effect diagrams, and program evaluation and review technique (PERT) analysis.

Decision trees use a flowchart structure to illustrate decisions and their possible consequences. These diagrams are drawn from left to right using decision rules in which the outcome and the conditions form a conjunction using an "if" clause. Generally, the rule has the following form:

if condition 1 *and* condition 2 *and* condition 3 *then* outcome, see Figure 1.6.

A strengths, weaknesses, opportunities, and threats (SWOT) analysis is a structured approach to evaluating a product, process, or service in terms of internal and external factors that can be favorable or unfavorable to their success. Strengths and opportunities (SO) ask "How can you use your strengths to take advantage of these opportunities?" Strengths and threats (ST) ask "How can you take advantage of your strengths to avoid real and potential threats?" Weaknesses and opportunities (WO) ask "How can you use your opportunities to overcome the weaknesses you are experiencing?" Weaknesses and threats (WT) ask "How can you minimize your weaknesses and avoid threats?" An example of a SWOT analysis is shown in Figure 6.3.

Risks can be assessed using a risk assessment matrix (Figure 4.4). The risk impact assessment is a process of assessing the probabilities and consequences of risks if they occur. The results of a risk impact assessment will prioritize ranking from most to least critical with importance. Risks can be ranked in terms of their criticality to provide guidance where

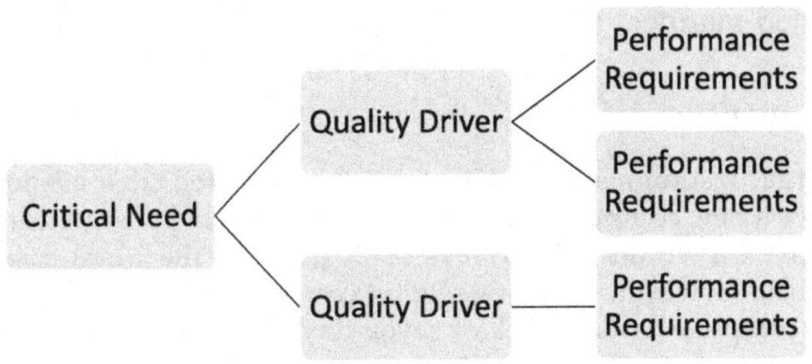

FIGURE 1.6
Tree diagram critical needs.

resources should be deployed or mitigation practices should be put in place in order to realize the risk events.

After a process is mapped, the cause and effect (C&E) diagram can be completed. This process is so important because it provides for root cause analysis. The basis behind root cause analysis is to ask "Why?" five times in order to get to the actual root cause. Many times problems are "band-aided" in order to fix the top-level problem, but the actual problem itself is not addressed.

The cause and effect diagram is also referred to as a fishbone diagram because it visually looks like a fish where the bones are the causes and the fish head is the effect, as shown in Figure 5.5.

The fishbone is general broken out to the most important categories in a system, which includes measurements, material, personnel (manpower), environment (mother nature), methods, and machines. These are also commonly referred to as the 6Ms. In addition, the headers can be named by performing an affinity diagram with the brainstormed ideas and using the theme for each grouping as the header for each fishbone. This process requires a team to do a great deal of brainstorming where they focus on the causes of the problems based on the categories. The "fish head" is the problem statement.

Program evaluation and review technique (PERT) analysis is an evaluation method that can be applied to time or cost. PERT provides a weighted assessment of time or cost. There are three key parameters for each scenario: optimistic (O) or best case scenario, pessimistic (P) or worst case scenario, and most likely (ML) scenario. From these parameters, the expected time, T_E, is calculated as in Equation 1.1:

$$T_E = \frac{O + 4M + P}{6} \tag{1.1}$$

The expected time calculation provides for a heavier weighting of the most likely scenario but also considers the best and worst cases. If the best or worst case scenario is an extreme situation, the PERT will account for it in a weighted manner.

EXAMPLE

An organization needs to determine the expected time needed to receive a new computer. The supplier generally over estimates and has given a window of 3 weeks or 21 days for the worst case (P) scenario. However, you have several coworkers that have received their computer in one week or seven days. This would be the best case (O) scenario. The most likely scenario is the median of supplier's delivery time, which is ten days for ML. Calculate the expected time.

$$T_E = \frac{O + 4M + P}{6}$$

$$T_E = \frac{7 + 4(10) + 21}{6}$$

$$T_E = 11.33 \text{ days}$$

Lean Philosophy as an Enterprise Solution

Conclusion

A Lean enterprise will improve quality, reduce lead time, and reduce costs.

The main process for a Lean enterprise is to map the value stream, identify the waste, create the future state, and continuously improve. The concept of Lean is to eliminate any type of waste possible while incurring zero defects. Lean always considers customer demand and their flexibility and encompasses a system that is able to adhere to these concepts.

Questions

1. What is status quo?
2. Why is change management important in launching a Design for Six Sigma (DFSS) initiative?
3. Why do people fear change? Describe a situation that you have experience where you had to lead change. What resistance did you encounter?
4. Why is it important to address "what's in it for me?"
5. Describe three ways you can reduce resistance to change.
6. What are the required elements for change?
7. Describe three strategies for overcoming resistance. Provide an example scenario on how you would use those three strategies.
8. What are the success factors with implementing DFSS?
9. Describe five strategies for effectively communicating DFSS. Provide an example scenario on how you would use those five communication strategies.
10. Describe the five qualities that show it is time for a change.
11. An organization needs to determine the expected time needed to design a new component. Based on past experience for a

similar component, the longest time was 185 days, the average time was 127 days, and the shortest time was 73 days. Calculate the expected time.

References

Heller, R. (1999). *Managing change*. Dorling Kindersley Ltd.

Recardo, R. (1995). Overcoming resistance to change. *National Productivity Review*, 14, 5–12. 10.1002/npr.4040140203

2

Management Support

Some are born into leadership, some achieve leadership and some have leadership thrust upon them.

—William Shakespeare

Lean Introduction

William Shakespeare said, "Some are born into leadership, some achieve leadership and some have leadership thrust upon them." Based on "Some are born great, some achieve greatness, and others have greatness thrust upon them," management must set the tone, example, and leadership for project execution. As the opening quote above opines, management practices must be above reproach. For effective and sustainable project control, management must create an environment for continual monitoring and mentoring of tasks.

Mentoring, in this case, relates to providing support and guidance for project execution. Management must provide leadership to take actions when needed for project control purposes. Management support is crucial for any type of project in order for successful results. Top-down approaches are mostly results oriented because people believe in the changes if management believes in the changes.

A deep commitment from the highest levels of management is necessary for the trust to occur. Senior management not only has to be committed but has to drive the process throughout the entire organization. This includes selection of projects based on data-driven exercises.

There will be changes in the tradition of how things are normally done. The implementation strategy consists of top management support and participation, project identification, resource allocation, data-based decision-making techniques, and finally measurement and feedback.

Making sure people are comfortable with making changes is a key component. The changes should also be able to be sustained so that "the old way of doing things" doesn't happen to occur in the future. Successful projects have managers that make projects their top priority devoting

DOI: 10.1201/9780429184123-2

their time, energy, and resources. Challenging employees in a productive way also makes a successful leader. Encouragement during the project from beginning to completion is absolutely a must.

The part of the control phase is where an ongoing support of the process occurs. Oftentimes Standard Operating Procedures (SOPs) are put in place along with a monitoring plan to ensure sustainability. Justifying improvements as a philosophy of management is a key component of project control. The process should be known as strategies for the project versus rules that must be adhered to.

The knowledge of systems is a key measure for stability in project controls. A system is a network of interdependent components that work together to accomplish a goal. These systems need to be management's responsibility.

It is management's responsibility not only to own the system but also to manage the system over time and keep up with improvements or optimizations. The system variations found in common causes are management's daily focuses with action to be taken on the systems. It is management's responsibility to own and optimize the entire system over time.

Proper management support for project control helps employees feel comfortable and as though they are making proper decisions themselves. The effects are everlasting and they as employees use it as a top-down approach and continue the momentum with their own team. The impacts that management makes on the team affect the business as a whole.

Specific examples are crucial for the impacts to be understood by the teams. It is important for the employees to not think of management as "mean" or the only decision-makers. Employees need their hearts and minds engaged and management needs to ask them frequently what they can do to help with that engagement. Management should see themselves as not the power players, but the top-down support that has special leadership techniques to influence others.

Communication is the main product of management. The communication needs to go both ways where the employees are able to confide in management about difficult situations and where management can explain expectations and timelines. Communication is also helpful to boost positive morale throughout the organization.

Positive reinforcement helps maintain structure and commitment. The reinforcement also helps build their egos and is noticed by others which in turn can be contagious.

Tricky situations for management are when cost-cutting initiatives need to take place in order for the business to be successful. This is complex for employees, but if demonstrated properly, can be understood. Direction on where the company is going with the initiatives will show people the reason for the cutbacks. Explaining that the culture will remain positive through the initiatives will keep the employees secure.

Most importantly, the participation of senior management using a top-down technique will provide for successful and operational organizations and project controls to be sustained.

Management's improvements are what distinguish quality management techniques when understanding Lean.

Project management is a key improvement tool for management to support. This includes helping other project managers by supporting their projects. This is the control portion that is critical for projects to be successful. Project management at this level does not necessarily mean that top management needs to participate in each team meeting, but is more of a support structure needed for employees to be motivated, remove obstacles, and show responsibility.

Management support includes prioritizing the projects through not only management's time but also by devoting key employees to projects that can have key effects on the business. Management support also includes basic common sense techniques such as encouragement, understanding of processes, and interpretation of particular methods or project analyses.

Roles and Responsibilities

- Select team members
- Allocate Resources Business Quality Council and the Quality Leader
  - Responsible for all strategic elements of Six Sigma from process management creation to sustaining Six Sigma as a cultural phenomenon
- Sponsor of the team – the Project Champion
  - Initiate the formation of the project
  - Responsible for establishing the preliminary impact on the business case
  - Provide the resources for the team to get their job done
  - Remove roadblocks

Responsibilities of the Champion

- *Business Quality Council and the Quality Leader*
  - Responsible for all strategic elements of Lean Six Sigma from process management creation to sustaining Lean Six Sigma as a cultural phenomenon

- *Sponsor of the team – the Project Champion*
    - Initiate the formation of the project
    - Responsible for establishing the preliminary impact on the business case
    - Provide the resources for the team to get their job done
    - Remove roadblocks

Responsibilities of the Team Leader

- Project Manager
- Sets agendas
- Facilitates each team meeting
- Keeps team on track meeting milestones
- Meets regularly with champion
- Maintains team effectiveness
- Leads implementation of solutions
- Hands-off finished project to Champion

Responsibilities of the Team Member

- Subject-matter experts
- Expected to do the work of DMAIC

Responsibilities of the Quality Leader

- Sets the strategic direction of Six Sigma within the organization
- Creates and maintains the Business Process Management system
- Works to obtain and sustain management involvement in Six Sigma
- Manages project tactics

Responsibilities of the Lean and Six Sigma Teams

- Black Belt is a full-time job. Green Belt is a part-time job (with regard to Six Sigma)
- Master Black Belts
  - Internal consultant
  - Coaches teams on tools and techniques
  - Facilitates a team when needed

The DMAIC Improvement Methodology

- Each step in the DMAIC model has what we call *Toll-Gates,* and there are usually two, three, or four *Toll-Gates* in each step.
- The DMAIC elements are similar to these *Toll-Gates.*
- *Toll-Gates* should be formally reviewed by a Project Champion before going forward.
- The arrows show that this model is not linear. Teams may return to Define during Data Collection in "the Measure phase" (Figure 2.1).

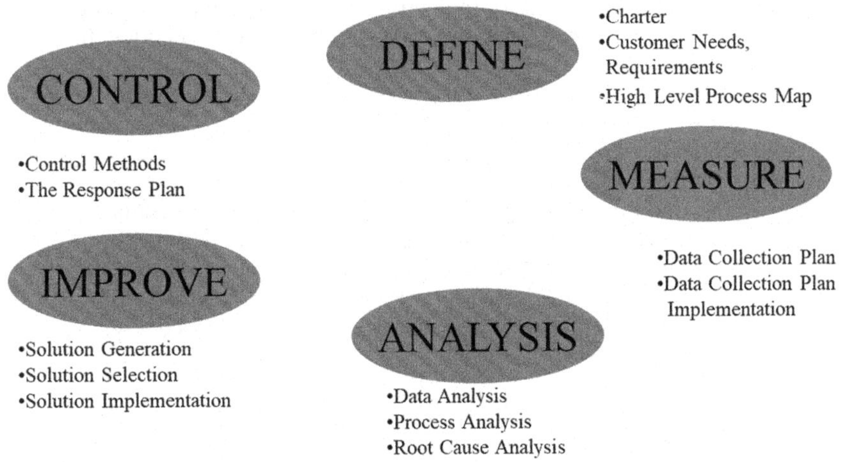

FIGURE 2.1
The DMAIC improvement methodology.

Project Resource Management Planning

Rarely do projects exist by themselves. They interact with other projects. There is competition between and within projects for limited resources and management attention. This will be a test of your negotiation and communication skills.

Resource management planning is a system for identifying and planning the resources needed to complete all projects within a specific time frame.

EXAMPLE

If your project portfolio is scoped and timed to align with your annual strategic planning review cycle, then you must plan resource needs for that same time frame. This is a common practice.

The plan is created during the initiation phase of the project life cycle, regardless of the type of project (planning implementations, problem-solving, any project requiring resources). Require each project team lead to prepare a resource management plan. There are templates out there that can be customized to facilitate this process. You may need to do some tool training.

Construct in tandem with the Scope of Work (SOW) document. The SOW feeds the plan.

Both documents should be submitted along with the project plan executive summary during the project prioritization process.

Master Black Belts must establish criteria for prioritizing Resource Management Plans. Consult Human Resources (HR) and finance for input. This is a great tool for budgeting and forecasting project costs and getting buy-in from Sponsors and Champions, not to mention the folks in finance.

This plan is a detailed and comprehensive three-step strategy:

1. List the required resources
    - Human
    - Identify all the roles involved in the project, full-time, part-time, and contracting roles. This is more than the team member lists. This includes all suppliers of inputs to the targeted processes. Process workers needed during the execution and some aspects of the control phases of the project life cycle
2. Equipment
    - Identify all of the equipment needed for the project: office equipment (computers, scan/fax/printers, mobile phones), telecommunications equipment (cabling, switches), and machinery (heavy and light, lasers, ovens, robotic assemblers)

3. Materials
    - Identify all of the non-consumable materials needed to complete project activities: office materials (copy paper, flash drives, ink cartridges) and materials required to build physical deliverables [wood, steel, and plastic (car); yarn, backing and laminate (carpet), flour, oil, and particulates]
    - Quantify each resource required
    - Number of people – Again not just team members, but process staff, suppliers, end users, especially people involved in the testing and validation of solutions, list the skills and experience specifications for each role
    - Number of each type of equipment
    - List the specifications for each type of equipment and its components
    - Number of each type of materials
    - List the specifications of each item of material required

If by now you're thinking this is a lot of work, I assure you it is. This information will help with cost-benefit analysis because during the project there will be times when the process will be off-line and not producing value for the customer. You need to capture those costs for the benefits/costs analysis, otherwise your return on investment (ROI) estimates will be significantly biased.

Construct a detailed resource schedule.

Now you are ready to pull all this together into a comprehensive schedule.

- You know the resources required to complete the project.
- You can estimate time frames for the consumption of each resource.
- You can track the quantity of each resource required per project per time frame.
- You can estimate total quantity of resources consumed per project per time frame (weekly, monthly).
- Assumptions and constraints can be identified within and across projects.
- There is so much you can do with this document, and the good news is that there are software packages that have these functionalities built in.

The objectives of Lean leaders are to do the following:

- Complete all projects to achieve overall organizational goals
- Evaluate long- and short-term priorities to guide decisions regarding resource allocations

- Acquire and maintain adequate resource levels
- Integrate project requirements with daily operations
- Balance development of management systems to support changing needs of projects by providing professional support to the people managing the projects

Earlier it was mentioned about the need to balance resources across the portfolio. What do you do if resources are constrained to the point of threatening project failure?

- Decrease the number of projects in the portfolio
- Reduce projects per team
- Build in extra capacity

Whatever approach you choose, use the data from the resource management plan (RMP) to justify your decisions.

A RACI matrix can help with project resource planning

- Any project needs a declared leader or someone who is responsible for the project's execution and success.
- You may hear references to RACI throughout your Six Sigma journey.
- RACI stands for Responsible, Accountable, Consulted, and Informed and identifies the people that play those roles.
- Every project must have declared leaders indicating who is responsible and who is accountable.

- **R – Responsible** Those who do the work to achieve the task. There is at least one role with a participation type of *responsible,* although others can be delegated to assist in the work required.
- **A – Accountable** The one ultimately answerable for the correct and thorough completion of the deliverable or task, and the one who delegates the work to those *responsible.* In other words, an *accountable* must sign off (approve) on work that *responsible* provides. There **must** be only one *accountable* specified for each task or deliverable.
- **C – Consulted** Those whose opinions are sought, typically subject-matter experts, and with whom there is two-way communication.
- **I – Informed** Those who are kept up to date on progress, often only on completion of the task or deliverable, and with whom there is just one-way communication.

Accountability Process

Purpose of Daily Accountability

1. What it IS
   a. A means of measuring performance against expectations
   b. A process for initiating action at the appropriate level to address deficiencies of each and every shift
   c. A process for asking for help once teams have attempted to (and were unsuccessful) address causes for missing expectations
   d. A tool to monitor and manage the current process to meet expectations
2. What it IS NOT
   a. A process for making structural or strategic changes to the process
   b. A cure all

Structure of Daily Accountability

Daily Accountability is structured with three layers. Each represents teams with different abilities and responsibilities.

1. Tier 1 (the Operating Team)
   a. The Tier 1 team is the production team. Let's suppose we have three or four shifts that comprise the production team. The production team includes shift maintenance and the Shift Supervisor.
   b. The production team is the group most directly engaged with the production process and thereby is generally the most knowledgeable of issues and most capable of quickly responding to issues.
   c. The production team is responsible for monitoring the production process and measuring performance relative to established goals. Information is posted on the Tier 1 board and various forms.
   d. The production team should identify the areas where performance is below expectations (misses).
   e. The production team should attempt to resolve any misses they experience and provide a team response on the Tier 1 boards. Any metric in RED (a miss) on a Tier 1 board requires a team response.

f. The production team should escalate any unresolved miss to the next tier.

g. The Team Leader should coordinate production team activities with respect to monitoring and reporting process performance relative to expectation, and posting team responses and escalation needs. The Team Leader ensures there is an appropriate hands-off approach at shift change.

EXAMPLES

- The Coating Area Rover serves as the Team Leader for the Coating Area (Feeder through #3 cooler or #3 oven).
- The saw operators alternate serving as the Team Leader for the Saw Area (Saws through Stackers, including rework).
- The Shift Supervisor will serve as the Team Leader for the packaging process.
    - The Tier 1 team is responsible for maintaining the Tier 1 board.
    - The Shift Supervisor works directly with the team to address misses and determine when escalation is warranted. The Shift Supervisor ensures there is an appropriate hands-off approach at shift change.
    - The Tier 1 teams must attempt to resolve problems. If they are not able to do so, they should provide sufficient details of the issue and request support from the Tier 2 team through escalation.
    - The Shift Supervisor should fill in for the Team Leader as needed.

1. Step-Up Supervisors support the production team in the same manner as the Shift Supervisor.
2. Tier 2 (the Support Group)
    a. The Tier 2 team consists of department operations and maintenance leadership and department engineering and quality support.
    b. It is sometimes referred to as Tier 2/3.
    c. The Support Group is responsible for the performance of the department.
    d. The Tier 2 team is responsible for any items escalated by the Tier 1 teams.
    e. The Tier 2 team should always provide a response to the Tier 1 teams.
    f. The Tier 2 team should attempt to resolve the issue. If they can't, they escalate the issue to Tier 4 through the escalation process.
    g. The Tier 2 team will address issues as quickly as they can. In some cases, this may be the same day the issue is

escalated. However, some issues can't be addressed that quickly. In those causes, the Support Group is expected to keep the production team updated on progress by posting on the Tier 1B board.

3. Tier 4 (Plant Leadership Team)
   a. The Tier 4 team is ultimately responsible for the performance of the plant.
   b. Issues that can't be addressed by Tier 1 or Tier 2 teams should be escalated to the Tier 4 team.
   c. The Tier 4 team is responsible for the issue and is also required to provide feedback on resolution back through the escalation process.

Daily Accountability Meetings

Information sharing is an essential part of daily accountability.

1. Team Meeting Expectations
   a. Team meetings are routinely conducted for team members to receive and provide information.
   b. It is difficult for all members of a primary production team to meet at once during the shift. Therefore, the Tier 1 level meeting is conducted at two locations to balance the need to monitor the production process and information exchange.
   c. It can be difficult to meet at the ideal time when there are major production upsets or line startup events. When these conditions exist at the beginning of the shift, the Team Leader should work with the Shift Supervisor to conduct the Tier 1A and Tier 1B meetings for that line at some point during the shift.
   d. Obviously the Shift Supervisor can't attend all Tier 1 meetings for all lines. The Shift Supervisor should attend a Tier 1A and Tier 1B meeting each shift. When on day shift, the Shift Supervisor should attend at least one Tier 2 meeting.
   e. Tier boards are maintained as shown in Figure 2.2:
2. Team Huddle (Team Meeting) Expectations
   a. Be on time
   b. No interruptions or distractions (focus on the huddle)
   c. No food (focus on the huddle)
   d. All participate
   e. Quickly work through the board focusing on misses or issues
   f. Provide actionable detail about issues escalated
   g. The team is expected to stand during the huddle
   h. Show respect for others and their ideas

Tier Board Name	Location	Leader	Ideal Time
CC 1 Tier 1A	Booth 2	Coating Rover	5 minutes after shift change
CC 1 Tier 1B	Saw Booth	Floor Saw Op	20 minutes after shift change
CC 1 Tier 2	Saw Booth	Area Supervisor	9:00 AM Mon – Fri
CC 2 Tier 1A	Booth 2	Coating Rover	5 minutes after shift change
CC 2 Tier 1B	Stackers	Floor Saw Op	20 minutes after shift change
CC 2 Tier 2	Stackers	Area Supervisor	9:15 AM Mon – Fri
Plant Tier 4	Exterior Press	Plant Manager	10:00 AM Mon – Fri

FIGURE 2.2
Tier board layout example.

The Meeting Rhythm (Working through the Tier Board)

1. People – Covers items specific to the employees assigned to the crew
   a. Tier 1 Meeting – The Team Leader should post their name on the placard. This is updated each shift. Target team huddle time is posted on the page and should be the time of the meeting unless disruptions in lines operation prevent it. The Supervisor should be notified if the team huddle can't take place as scheduled. The Tier 1A meeting focuses on the process from feeder through coating process. The Tier 1B meeting focuses on the process from saws through stackers, including rework. Both feeder ops, both booth ops, and the coating rover participant in the Tier 1A huddle. The stacker ops, rework ops, and one saw op participant in the Tier 1B huddle. Once the process is underway, the rework ops should alternate with the graders to allow participation.
   b. Announcements – Serves as a means of two-way communication. The Supervisor and Tier 2 members can provide general information to the team. The Team Leader and team members can post information for the Supervisor and Tier 2 members. Upcoming trial work and special scheduling events are good examples. The Team Leader can also use this space to provide reminders for the team.

c. Recognition – Team Leaders, Supervisors, and Tier 2 members can use this space to recognize a special achievement by teams or team members.
d. Line Up – Use this area to post current Line Up and personnel assignments. Also post work (break) schedule. Refer to as needed.

2. Performance
   a. Process performance is categorized by SQDIP. The acronym represents Safety, Quality, Delivery, Inventory, and Productivity.
   b. Safety – Safety will contain two areas of focus (Housekeeping and Safety Issues).
      i. Housekeeping – The line is divided into six zones. A zone will be audited each shift such that the entire process is audited every two days. The Team Leader will conduct the audit during the shift and post the results on both Tier 1 boards. The Team Leader should initiate action to address any scores in red on the Housekeeping Audit Results chart. That action represents the team response and should be noted in the Team Response space. Guidance on how to score each zone is provided in the Housekeeping Audit Plan. Any deficiencies the team can't address should be posted in detail on the Safety Suggestions and Issue Escalation Form.
      ii. Safety Issues – Any safety issues that arise during the shift should be posted on the Safety Issues Chart. The team should attempt to address any safety issue that arises. Team actions should be noted in the Team Response space. Any issues the team can't address should be posted in detail on the Safety Suggestions and Issue Escalation Form.
   c. Quality – Quality contains two areas of focus (Yield and Certain Line Checks)
      i. Yield – Near the end of the shift, the Graders and Rework Operators will notify the Saw Booth of the quantity of culls generated and the reason for the culls. The Saw Booth Operator will calculate the yield for the shift. The Team Leaders will post the results on the Yield Charts on both Tier 1 boards. If the yield result is in red, the Team Leaders should note the team's efforts to address the causes in the Team Response space. Any issues the team can't address

should be posted in detail on the Quality Suggestions and Issue Escalation Form.

ii. Line Checks – Results of certain line checks will be posted on a Tier 1 board. This does not mean that other current line checks are not to be completed. Line Checks included on the Tier 1 boards are currently areas of special attention.

1. Tier 1A Line Checks – Paint solid check results should be the current posted line check. During normal operation, there are eight solid checks per shift. The Team Leader should post the percentage of line checks in specification on the Tier 1A board. If the percent "in spec" falls in red, the Team Leader should post the actions initiated by the team to improve the percentage "in spec." This should be posted in the Team Response space. Any issues the team can't address should be posted in detail on the Quality Suggestions and Issue Escalation Form.
2. Tier 1B Line Checks – Number of Loads placed "On Hold" should be the current posted line check. The Team Leader should post the results on the Tier 1B board. If the number of Load placed "On Hold" exceeds the acceptable number and falls in red, the Team Leader should post the actions initiated by the team to reduce the rate at which loads are placed "On Hold." This should be posted in the Team Response space. Any issues the team can't address should be posted in detail on the Quality Suggestions and Issue Escalation Form.

   a. Delivery – Delivery represents the amount of product produced during the shift relative to plan. The result is displayed as skins produced. The Team Leader posts the number of skins produced during the shift on the Delivery Chart. If the result is in red, the Team Leader should explain what actions were initiated by the team to address the miss. In some cases, the miss is caused by scheduled downtime such as scheduled die changes. Use the Team Response space for this posting. Any issues the team can't address should be posted in detail on the Delivery Suggestions and Issue Escalation Form.
   b. Inventory – Inventory is not controlled by operations. Therefore, the team is not responsible for reporting on it. However, it is important that the team understand that the plant is measured on inventory management. Tier 1 and 2 teams will refer to site-level results posted monthly on the scorecard.

   c. Productivity – Productivity is a function of uptime, speed, and yield. We refer to these as utilization, efficiency, and yield. Standards exist for each production line. The product of multiplying by 3 is called Net Productivity Index. These should be calculated as is currently done and posted on both Tier 1 boards by the Team Leader. Productivity performance should track delivery performance. Any results in red can be explained as part of the Team Response to a delivery or quality miss. No Team Response space or Suggestions and Issue Escalation Form is provided for productivity.
3. Tier 1A and Tier 1B Board Posting
   a. Much of the information on the two boards is the same. Again, two boards are used so all employees can participate in meetings during the shift while the line is operating.
   b. The Support Group will meet at the Tier 1B board. Therefore, information specific to the Tier 1A board (Team Responses and Escalation Issues and Details) should be transferred to the Tier 1B board.
   c. The Support Group will post its updates on the Tier 1B board. These should be transferred to the Tier 1A board at some point during the shift, so team members meeting in the control room can be made aware of the feedback provided.
4. Site-Level Metrics – Site or Plan Metrics are posted on each Tier Board. The information is updated monthly. This allows everyone to see how the performance of a production line is linked to overall plant performance.

The Lean Management Operating System consists of the following:

- Daily Management/Continuous Improvement
- Strategy Deployment

Culture is 80%, while tools are 20% of the results of a Lean conversion (Figure 2.3).

To change culture, we change the Experiences, Management System, Behaviors, Tools, Practices.

The Lean leader cultural transformation consists of several management styles as shown in Figure 2.4.

In order to lead in a Lean environment, it is important to ensure the following aspects are met:

- Build the Vision – Considerable effort is required to understand Lean and then to interpret the underlying principles and practices as they would apply to your company.

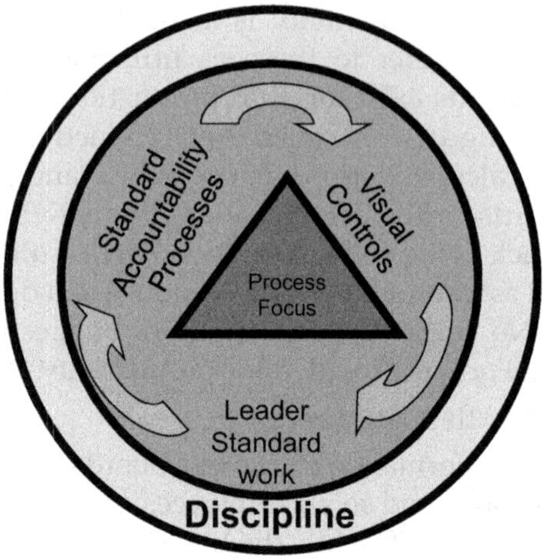

FIGURE 2.3
Discipline.

- Need Establishment – Few companies are willing to make the dramatic, pervasive changes required unless they are experiencing a major challenge or even a threat of surviving the big picture.
- Foster Lean Learning – Essentially all key leaders need to be brought up to speed on Lean. Regular, frequent meetings need to be organized.

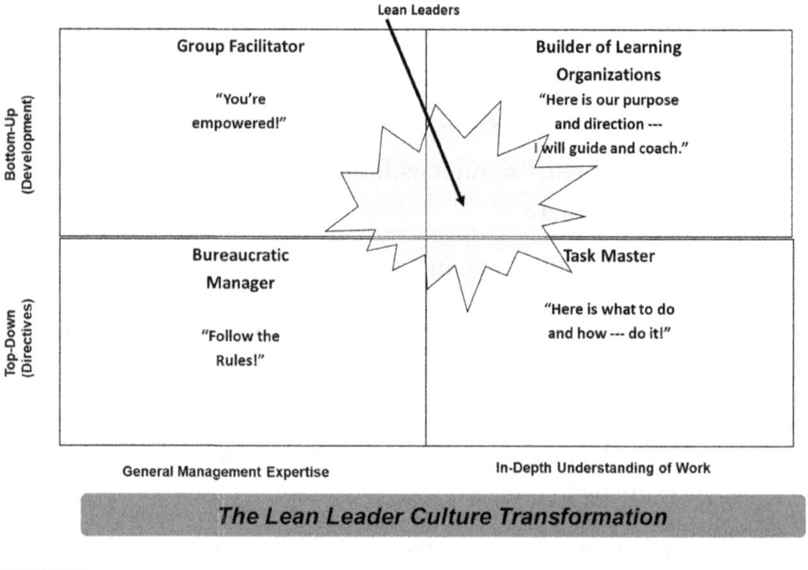

FIGURE 2.4
The Lean leader culture transformation.

- Commitment – There should be no ambiguity regarding decision authority or resource control relative to the decision to pursue a Lean transformation.
- Your either in or out … sitting on the sideline won't work!
- Non-negotiable(s) – The decision to pursue Lean, once made, must be viewed as non-negotiable and irrevocable. Full buy-in of all senior managers is mandatory. Expectations of each manager must be made clear.

When leading in a Lean environment:

- Over communicate!
- The deployment of a Formal Lean Enterprise Transformation represents a major organizational change.
- Long-standing operational elements that are deeply ingrained could be radically altered or eliminated.
- **Unwavering leadership commitment** may be the most critical component of success in managing in a Lean environment.
- Active and **visible** executive sponsorship is the number 1 contributor to change management success.
- Lack of commitment will be immediately visible to the team and must be addressed.

The communication of strategies to employees early on is key in order for trust to be established. Fluid strategies reflecting today's dynamic business conditions can make communicating business objectives as difficult as writing on water. Yet a workforce that better understands a Lean strategy is better able to execute it. Credible leaders convey clear, consistent messaging that links every employee to the Lean strategy and drives engagement, productivity, and success. They develop a comprehensive communication plan to carry employees at all levels from awareness and understanding to commitment and enhanced performance. They deliver messages in a clear and compelling manner and monitor the effectiveness of the delivery and its impact on progress toward "True North."

The only way to move a culture is to change the experiences people have in the organization.

People will tend to embrace change when they have a role to play and are unlikely to embrace change willingly if they are seen merely as a change target. It is therefore imperative that they are engaged up front to ensure that the elements are well built and to ensure that there will be commitment to make them work. There are three keys – Strategic Management, Daily Management, and Talent Management (Figure 2.5).

FIGURE 2.5
Three keys.

Create a culture reflecting the values with which employees can identify. An organization's culture has a life of its own. No leader or group of leaders can control it entirely. However, leaders can influence a culture in ways that will drive engagement. The most effective leadership behavior is leaders showing that they value employees. Credible leaders must do more than simply say they value employee contribution. They need to make it real and demonstrate it. To maximize employee contribution, top management must be aligned and show employees that they really count. This includes their ideas, contributions, values, and commitment. It has to be ongoing, not an annual event. An organization's culture has a life of its own. No leader or group of leaders can control it entirely. However, leaders can influence a culture in ways that will drive engagement. The most effective leadership behavior is leaders showing that they value employees. Credible leaders must do more than simply say they value employee contribution. They need to make it real and demonstrate it. To maximize employee contribution, top management must be aligned and show employees that they really count. This includes their ideas, contributions, values, and commitment. It has to be ongoing, not an annual event.

Leading in a Lean environment has a lot of strategy such as:

- **How We Lead and Manage**
  - Leaders *own* process improvement – coach and mentor teams
- **How We Design Our Work**
  - Work designed to enable *simple* visual management

- **How We Improve**
    - "Standardized work" – basis for *continuous improvement*
    - Problems fixed as they occur – *and stay fixed*
- **How We Sustain Improvements**
    - *Strategy* Deployment Process and robust performance execution process

Implement the organization's strategy effectively by thinking effectively.

Mergers, acquisitions, restructurings, and dynamic market conditions all can necessitate strategic change for Lean. But to ensure a change in direction doesn't mean going off in all directions, the organization must maintain focus. Credible leaders execute strategic change by first understanding the complexities of the change and facing the realities of the external forces putting pressure on their business. Then, they optimize the organization's structure, capacities, and capabilities; its people, systems, and processes; and its leadership performance. They prioritize critical issues, analyze root causes of performance inhibitors, and identify capability gaps. Once strategy has been clearly articulated and agreed upon, they clarify fit-for-purpose structures and roles; deploy people, systems, and processes; and assign capable leadership at all levels. And finally, credible leaders measure for impact on actual business performance.

Managing complex change is difficult in an organization. Figure 2.6 represents the change aspects.

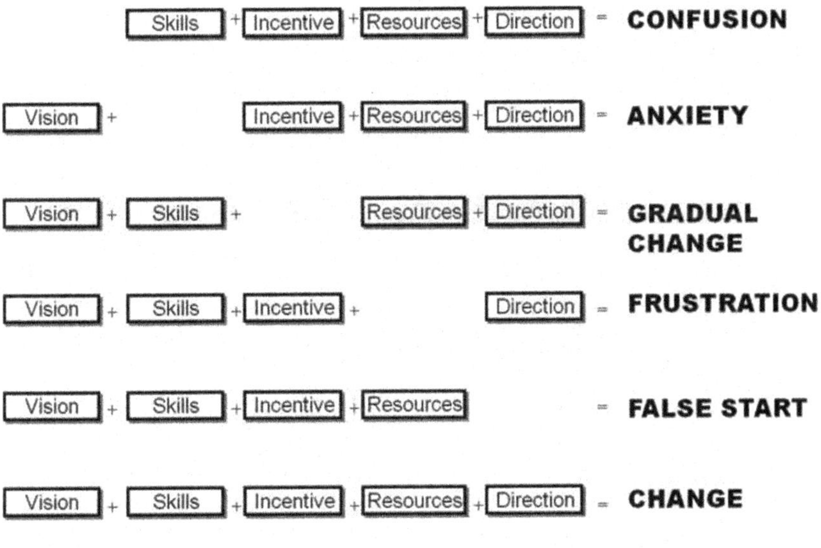

FIGURE 2.6
Change aspects.

The critical takeaways of Lean leadership consist of your role. As a leader, you are now the makers of leaders. Do not accept the unacceptable. Results, processes, and behaviors that are not desirable will never be accepted. Hold yourself to a higher standard and be a role model of behaviors and accountability. Hold others accountable: peers, subordinates, and superiors. Respectfully and with professionalism, but hold them accountable. Bring accountability to a personal level: visually in everything we do. Coach "on Gemba": Redirect, Recognize, Motivate on the spot. Continuous improvement: everyone, everywhere, everyday – it is a must for survival!!

Coaching

Why Coach?

- Employees at all levels accept ownership and accountability
- Employees develop a solutions focus
- Employees require less daily and direct supervision
- Engagement
- Retention
- Increased business performance
- Skill development at all levels
- Effective and efficient transfer of critical knowledge
- The success of one manager is exponentially multiplied
- Creation of a leadership pipeline (Figure 2.7)

Supporting Values	Inhibiting Values
Trust and openness	Mistrust and fear
Tolerance of mistakes, learning from them	Intolerance of mistakes, blaming of the perpetrator
Careful attention to hiring the right people	Lack of careful attention to hiring, credibility of members
Learning for the long term is important	We should focus single-mindedly on evaluating today's performance
Reward systems shouldn't punish time spent developing people	Reward systems focus only on short-term results
People are valued as individuals	People are a "means to an end"

FIGURE 2.7
Values.

You cannot coach someone who does not trust you and/or believe that you have their best interests at heart.

The Socratic coaching structure can be a direct reporting relationship or another internal coach. It should be used to understand (grasp the situation) as well as get the learner to reflect. Use during Gemba walks, tier meetings, etc.

There are six types of questions that Socrates asked his pupils.

- Conceptual clarification questions
- Probing rationale, reasons, and evidence
- Probing assumptions
- Questioning viewpoints and perspectives
- Probe implications and consequences
- Questions about the question

Ask the right questions:

- Open-ended questions
    - How does this help you?
    - Why is this our standard?
    - What would changing this do for you?
    - When do you use this equipment?
    - Who is a part of this process?
- Yes/No questions are No/No questions
    - Knee-jerk response/pattern of response
    - Little information/insight
    - No opportunity for the responder to show their knowledge
    - No opportunity to guide the thought process
    - Can be pedantic
    - You might miss the opportunity to learn facts that you did not know!

Scientific thinking is a specific type of thinking that makes you think of the following types of thinking:

- Problem-solving is a unique type of thinking.
- To help people develop skills in this area, Lean has a special type of questions.
- The questions are designed to create a habitual, scientific way of approaching a problem.

There are particular coaching process roles:

- LEARNER
    - Applies the improvement principles
    - Seeks to find improvements scientifically
    - Can apply Plan Do Check Act (PDCA)
- COACH
    - Ensures the learner is following the improvement kata (approaching improvement scientifically).
    - Conducts daily coaching cycles using the five coaching kata questions.
    - The coach guides, but does not do.
    - The coach is responsible for their learner's results.

Conclusion

Improvement and coaching are closely linked to strategy deployment. This does not mean the same method cannot be used for other, smaller goals. This type of coaching is very specific – it aims to teach people a structured and scientific way to solve problems and creates accountability to do something toward a goal every day. This is not developmental or performance coaching which should be done by someone's direct supervisor. You can expect that some people will not like this new shift. It was much easier to ask you the question and have you give them the answer. Thinking takes effort and expect frustration and annoyance; however, most importantly, don't give up!

3

Hoshin Kanri

Plans are nothing, planning is everything.

—Dwight D. Eisenhower

Lean and Strategic Thinking

The Japanese quality thinking began before 1645. Miyamoto Musashi wrote a guide to samurai warriors on strategy, tactics, and philosophy entitled *A Book of Five Rings*. Musashi was a Japanese swordsman who became legendary for his duels and distinctive style of swordsmanship. In his book, Musashi states, "If you are thoroughly conversant with strategy, you will recognize the enemy's intentions and have opportunities to win."

A corporation's strategic plan must be integrated with the macrolevel value stream mapping (VSM) to identify the optimal improvement opportunities. This promotes strategic thinking. Often improvement activities are identified with silo thinking. The effects on other systems or processes within the organization are not considered. Improvements in one area can have a negative impact on another business area.

Senior leadership including the CEO and directors should use Hoshin Kanri to develop long-term strategic objectives. Mid-level managers should then use macrolevel value stream mapping to identify the areas of improvement to achieve strategic goals. Finally, department teams should use Lean and Six Sigma tools for process improvement.

To think strategically, the senior leaders of a corporation – in other words, *you* – should first determine where it is going – the vision. Then you need to identify your business's key processes. Next, you should perform a gap analysis between your organization's current state and your vision. This will lead to a strategic approach to continuous improvement.

To become a Lean enterprise, you must integrate Lean throughout all levels of your organization. This means breaking down the silos and changing the focus of process improvement to a global perspective. What you need is a holistic approach to continuous improvement throughout

DOI: 10.1201/9780429184123-3

your corporation. This will enable your corporation to make improvements that get the biggest bang for the buck, rather than suboptimal improvements. Lean strategy deployment breaks down these barriers and enables a holistic approach to continuous improvement that links to the long-term goals of your corporation.

History of Hoshin Kanri

Hoshin Kanri began in Japan in the early 1960s as statistical process control (SPC) became total quality control (TQC). Hoshin Kanri is most commonly referred to as Policy Deployment (PD). "Hoshin" means *shining metal, compass,* or *pointing the direction*. "Kanri" means *management* or *control*. Here's an overview of what Policy Deployment is and does.

- Policy Deployment is a systems approach to management of change in critical business processes.
- It is a methodology to improve the performance of critical business processes to achieve strategic objectives.
- Policy Deployment improves focus, linkage, accountability, buy-in, communication, and involvement in a corporation.
- It links business goals to the entire organization, promotes breakthrough thinking, and focuses on processes (rather than tasks).
- Policy Deployment is also a disciplined process that starts with the vision of the organization to develop a three- to five-year business plan and then drives down to one-year objectives that are deployed to all business units for implementation and regular process review.

Policy Deployment is a business management system designed to achieve world-class excellence in customer satisfaction. The system, beginning with the voice of the customer, continuously strives to improve quality, delivery, and cost. The system provides the necessary tools to achieve specific business objectives with the involvement of all employees.

As shown in Figure 3.1, you should take the voice of the customer to drive your business targets. Then, using Policy Deployment as your management strategy, you should drive down this strategy throughout all levels of your business to focus on safety, quality, delivery, and cost. Then using foundational Lean Six Sigma tools such as pull, 5S, single minute exchange of dies (SMED), standard work, total productive maintenance (TPM), and value stream mapping (VSM), you can focus on continuous

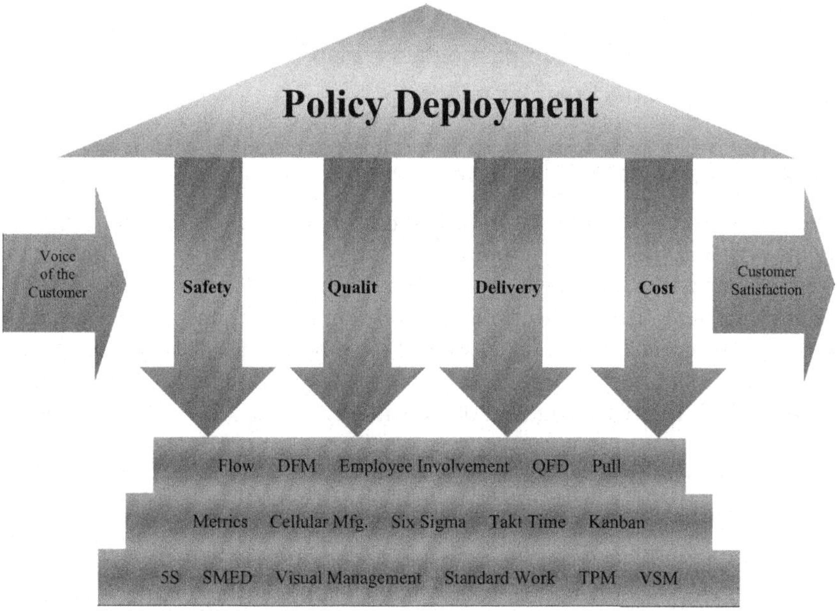

FIGURE 3.1
Strategic business system.

improvement. This leads to improved customer satisfaction, which further leads to improved sales growth for your organization.

Two Levels of Policy Deployment

Policy Deployment is a methodology to capture strategic goals and integrate these goals with your entire organization's daily activities. The two levels of Policy Deployment include:

(1) Management or strategic planning
(2) Daily management

Planning for Policy Deployment

Effective planning is critical for the long-term success of a corporation. There are five main steps for effective planning:

- Step 1 – Identify your critical objectives
- Step 2 – Evaluate the constraints

- Step 3 – Establish performance measures
- Step 4 – Develop an implementation plan
- Step 5 – Conduct regular reviews

Daily Management of Policy Deployment

Daily management involves applying Plan-Do-Check-Act to daily incremental continuous improvement to identify broad system problems in your organization. Once you gain a breakthrough improvement in the system problem, then the improvement becomes the focus of daily continuous improvement activities. Hoshin Planning is the system that drives the continuous improvement and breakthroughs. Policy Deployment involves both the planning and deployment:

- Develop your targets
- Develop your action plans to achieve your targets
- Deploy both

The concept of hierarchy of needs was introduced by Maslow and it outlines the basic needs that must be met before moving on to a higher need. Maslow's hierarchy of needs is illustrated in Figure 3.2.

In concurrence with the hierarchy of need that must be met for an organization to move on to its higher need, there are five levels of organizational needs:

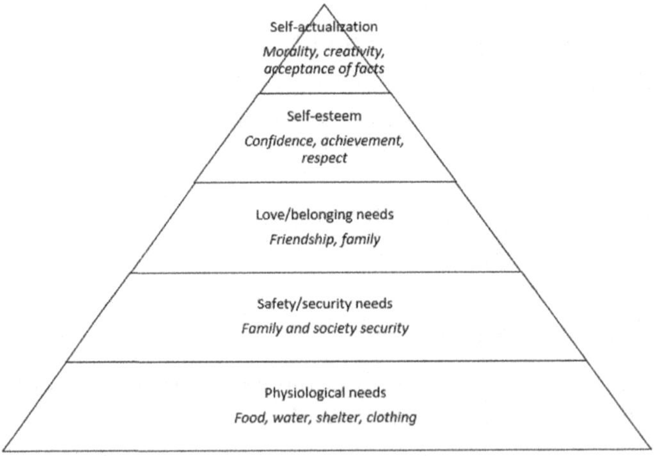

FIGURE 3.2
Maslow's hierarchy of needs.

- Level 1 – Core vision
- Level 2 – Alignment
- Level 3 – Self-diagnosis
- Level 4 – Process management
- Level 5 – Target focus

Figure 3.3 illustrates the five levels of organizational needs.

The five levels of organization needs are directly linked to the six Hoshin Planning steps and five Hoshin methods. Figure 3.4 shows the linkage between the organizational needs, Hoshin Planning steps, and Hoshin methods.

Within the Hoshin Planning process, there are six main steps. Each of these steps is discussed in the following paragraphs.

Step 1: Develop a five-year vision – Top management should develop a five-year vision to define the strategic objectives of your organization based on the internal, external, and environmental

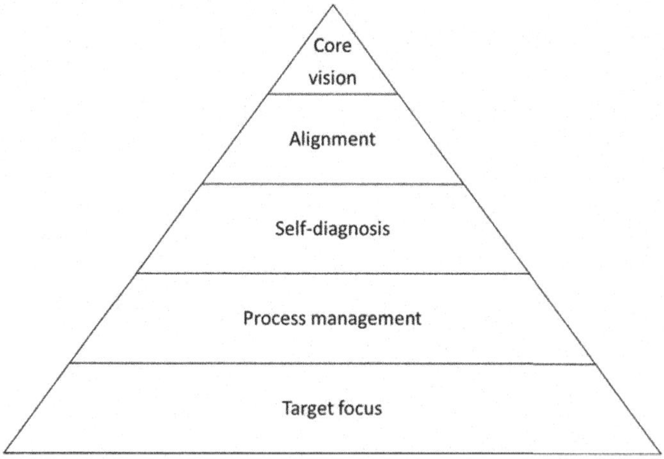

FIGURE 3.3
Levels of organizational needs.

Organizational needs	Hoshin planning steps	Hoshin methods
Core vision	5-year vision 1-year plan	Hoshin strategic plan summary
Alignment	Deployment	Hoshin plan summary
Self-diagnosis	Implementation	Hoshin action plan
Process management	Monthly reviews	Hoshin implementation plan
Target focus	Annual review	Hoshin implementation review

FIGURE 3.4
Linkage of organizational needs.

challenges your organization faces. The five-year vision is the future target for your organization. It is defined by your organization's goals, capabilities, and culture. It is a statement of where your organization wants to be in the future. The vision is a communication tool for senior leadership to relate the ideal future for the organization. For example, Company XYZ has a vision to be one of the top ten companies in their field in five years.

Step 2: Develop a one-year plan – Based on your five-year vision, you should then develop a one-year plan to outline continuous improvement activities that will enable your company to achieve its long-term strategy. The one-year plan is linked to the five-year vision by taking an incremental step and defining key targets to attain that step. The purpose of this step is to focus the activities throughout all levels of your organization on addressing your external issues and improving your internal problems. As part of this step, you should analyze the external factors, including your competition and the economy as a whole. In addition, you should analyze past problems so that your organization does not repeat them. Top management must then prioritize the objectives based on safety, quality, delivery, and cost. Developing the one-year plan blazes the path for the company to achieve its five-year vision.

Step 3: Deploy your one-year plan – The next step is to deploy your one-year plan to all departments within your organization. Deploying your one-year plan is where you begin to set measurable goals for each department. This is a planning step to determine specific opportunities for improvements within each department. At this point, a strategy is set to achieve the path set by the one-year plan. This leads to Step 4, which is implementation.

Step 4: Implement continuous improvement activities – Each department must drive continuous improvement activities that are aligned with your one-year plan and five-year vision. This is a step where the improvement process begins. The prior Steps 1–3 involved planning for improvement, now you are performing the improvement activities. For example, an improvement activity in this step could be a SMED event in a cell to reduce changeover time. The improvement activities in Step 4 must tie directly back to the one-year plan. This involves developing a master plan with appropriate measures and goals.

Step 5: Conduct monthly reviews – You should track the progress of your continuous improvement activities using quantitative metrics and communicate them to senior leadership (CEO and Directors) in a monthly review. The monthly reviews should link directly to your deployment of the one-year plan. You should monitor your actual improvements against your planned improvements as a monthly self-diagnosis and to ensure the corrective actions are sustained. During a formal review to management, each department should present the tasks addressed, a problem analysis, and the problem-solving results. As the Hoshin Planning implementation becomes more engrained into your organization, the review may become more of a highlight or overview of the problems and corrective actions.

Step 6: Perform an annual review – Finally, you should conduct an annual review to monitor your progress and capture your organization's results. The annual review provides an opportunity to ensure that the implemented projects helped attain the one-year plan and five-year vision. This is a check of how your implementation affected the organizational metrics set out in your five-year vision. Based on the results of the previous year and the effectiveness of your implementations, a new one-year plan is developed to set the targets and goals for the upcoming year. In addition, the organization may redevelop their five-year vision based on the current business environment.

The strategic planning process is a critical piece of the puzzle as it lays the foundation for the next step when you will cascade your organization's strategic goals throughout your organization. In order to accomplish the six Hoshin phases, there are Hoshin methods that align with the phases including:

1. Hoshin Strategic Plan Summary
2. Hoshin Action Plan
3. Hoshin Implementation Plan
4. Hoshin Implementation Review

Hoshin Strategic Plan Summary

Effective planning is critical for creating an organizational strategy and vision. The next step is to cascade down the strategic vision and goals into all levels of the organization. The first step in Policy Deployment is to develop your strategic plan summary. The Hoshin Strategic Plan Summary links the strategic vision of the organization with measurable goals. The Hoshin Strategic Plan Summary shown in Figure 3.5 illustrates the relationship between an organization's strategic goals, core objectives,

HOSHIN STRATEGIC PLAN SUMMARY																					
					Core Objectives																
			●		Core objective 1	O				●			●	●	●	O	●			●
	●		O		Core objective 2					O			●	●		●	O	●		
		●		●	Core objective 3	●	●	O	●	●	O	●			●	●		●	O	●
Strategic Goals	Strategic goal 1	Strategic goal 2	Strategic goal 3	Strategic goal 4		Metric 1	Metric 2	Metric 3	Metric 4	Metric 5	Metric 6	Metric 7	Metric 8	Metric 9	Director of Operations	Director of New Business	Director of Marketing	Director of Engineering	Director of Quality	Director of Finance

FIGURE 3.5
Hoshin Strategic Plan Summary.

metrics, and ownership. A value stream map gives you a picture of all the activities required to produce a product. The Hoshin Strategic Plan Summary also provides a picture of the overall strategy of an organization and how the strategy cascades throughout all levels of the organization. The linkage is clear on how each strategic goal is measured and who has the ultimate responsibility.

How to Create a Hoshin Strategic Plan Summary for Your Organization

Let's look at each aspect of Figure 3.5 to illustrate how you can develop your own Hoshin Strategic Plan Summary.

List your strategic goals – In Figure 3.5, the strategic vision is listed vertically on the left-hand side under the heading "Strategic Goals." These are the broad strategic goals of the organization for the next five to ten years. The strategic goals should be what the organization needs to do to ensure long-term success.

List your core objectives – Next, we drive down a level to get more specific measurable goals. As you can see in Figure 3.5, these specific measurable long-term goals (for the next three to five years) are listed horizontally under the heading "Core Objectives."

Make sure your core objectives link to your strategic goals – To ensure these goals are linked to the overall strategic goals in the plan summary (in Figure 3.5), a coding system is used to show the strength of the linkage:

- A filled-in circle (●) indicates *a strong relationship* between the strategic goal and the core objective.
- An open circle (O) indicates *a direct relationship* between the strategic goal and the core objective, but the core objective is not necessarily one of the key drivers for that strategic goal.

The key reason for showing the relationships in the strategic plan summary is to ensure that you are adequately addressing the strategic goals of your organization through your core objectives and that you are adequately measuring them by using the appropriate metrics.

Identify who is responsible: identify your metrics – Finally, you also want to ensure that the proper ownership exists to drive the necessary improvements. You need to assign each improvement activity to a specific person for accountability.

Determine how you will measure improvements for meeting each goal – Now, you cascade the overall objectives into how you will manage the business and measure the progress of your process improvement activities. In this step, you now need to determine your metrics and short-term goals (for the next one to two years). These metrics will drive how you will manage your business and how you will prioritize your process improvement activities. List your metrics vertically on the right side of the matrix, as shown in Figure 3.5.

Make sure you tie your metrics to your core objectives. To ensure your organization will meet your measurable core objectives, these metrics must meet several criteria:

- First, they must be *measurable* (quantitative not qualitative). These metrics assess the effectiveness of your process improvement efforts.
- The metrics must also be *baselined* to show your current performance or *benchmarked* against your competitors or your industry's standards.
- In addition, these metrics must be *achievable*. When your metrics are unachievable (i.e., zero internal defects), your employees will become discouraged, and your system will fail. In contrast, if you set realistic goals (i.e., zero external defects), your employees will team together to ensure your success.

Next, use the same coding scheme:

- (●) to show strong relationships between your metrics and your core objectives

- (O) to show direct relationships between your metrics and your core objectives

Identify who is responsible for meeting your core objectives: assign ownership – The final step in the strategic plan summary is to assign ownership of the core objectives. The ownership at this level falls on your organization's executive leadership because they own the responsibility of creating the strategic vision and driving it down through your organization. In Figure 3.5, the leadership team is listed vertically at the far right of the matrix. Here, again, we use the coding scheme to illustrate ownership that shows the following:

- Ownership of a core objective with a filled-in circle (●)
- Cursory responsibility with an open circle (O)

Because these are tied into the strategic goals of your organization, there will be core objectives with several owners. For example, if one of your strategic goals is to drive new business through your product offerings, this will involve engineering and marketing (as well as possibly others). Therefore, in this example, both engineering and marketing would have the main ownership.

Hoshin Plan Summary

The Hoshin Plan Summary details the strategic goals and cascades them down to the department level. While the Hoshin Strategic Plan Summary was at the highest organization level, the Hoshin Plan Summary is the tactical plan for each department. Now you need to drive the strategy down to each department. Figure 3.6 illustrates the Hoshin Plan Summary.

The first column in Figure 3.6, labeled "Strategic Goals," should correspond to the strategic goals listed in your Hoshin Strategic Plan Summary (Figure 3.5). Now that you're moving down to the department level, the management owner will be the person in charge of that department. Each department will have its own Hoshin Plan Summary. In some cases, depending on the management structure of your organization, the management owner will be the same here as the core objective owner in your Hoshin Strategic Plan Summary.

The next two columns are for your short-term and long-term goals. Your long-term goals may, in some cases, correspond to the strategic goals from the left side of your Hoshin Strategic Plan Summary. However, your goals should definitely correspond to the metrics you outlined in the Hoshin Strategic Plan Summary. This ensures that you align the proper activities

Hoshin Plan Summary								
Strategic Goals	Management Owner	Goals		Implementation Strategies	Improvement Focus			
		Short-term	Long-term		Safety	Quality	Delivery	Cost
Strategic Goal 1								
Strategic Goal 2								
Strategic Goal 3								
Strategic Goal 4								
Strategic Goal 5								

FIGURE 3.6
Hoshin plan summary.

with your overall strategic vision. Based on the metrics you previously outlined, your short-term and long-term goals should already be developed. These may not necessarily be the same for each department.

For example, let's look at external defects, which are measured as external parts per million (PPM): a manufacturing department's short-term and long-term goals should be very aggressive. On the other hand, the engineering department should also be focusing on improving the product design using tools, such as design for manufacture and assembly, which will in turn reduce external defects. Therefore, engineering department's short-term and long-term goals for reducing external defects will not be as aggressive. Also, your organization as a whole must be aligned and managed by your senior leadership team to ensure that your various departments come together to provide the overall necessary reduction in external defects that your organization is seeking. This provides a common goal for multiple departments to work together to achieve and eliminates silos.

In addition, with respect to the goals, there may be several metrics that relate to a core objective. As noted in Figure 3.6, you may need to list your core objective in multiple rows to correspond to the appropriate metrics. The various metrics for a core objective may then call for a different implementation strategy. For example, if your core objective is to improve product quality, this can be measured with internal PPM and external PPM. Internal PPM may be handled with an internal Six Sigma project as the implementation strategy. On the other hand, the implementation strategy to reduce external PPM may be to implement a poka-yoke device. Therefore, you would want to list these in two different rows to highlight that they are two different metrics with different implementation strategies.

Implementation Strategies for Your Hoshin Plan Summary

The next step in developing your Hoshin Plan Summary is to develop your implementation strategies. This is critical in how your organization makes process improvements appropriately, using the most efficient and effective technique. Each department must develop a strategy on how it will achieve its short-term and long-term goals. The team members developing the strategy should revisit their current state map(s) to understand all the activities involved. This will enable them to select the most effective technique – whether it is a Six Sigma project, 5S, standard work, SMED, TPM, etc.

Decide Where You Want to Focus Your Improvement Efforts

The final step in completing your Hoshin Plan Summary is to determine your improvement focus. Typical focus areas for organizations include safety, quality, delivery, and cost. Here, again, use the same coding scheme:

- The filled-in circle (●) to show strong relationships between the implementation strategy and its impact on safety, quality, delivery, and cost
- The open circle (O) to show cursory relationships

The purpose of showing the relationships in the Hoshin Plan Summary (Figure 3.6) is slightly different. Here, you want to balance your improvement efforts. You still need to ensure that you link your implementation strategies to your improvement goals, which are linked to your core objectives. This common thread of linkage must be clear. But you also want to make sure that the implementation strategies you develop will have an impact on your improvement focus areas. For example, if you develop an implementation strategy that only impacts quality but does not impact safety, delivery, or cost, then that may be a signal that it is not the most effective strategy. You want an implementation strategy that impacts more than one focus area. As with anything, there must be a balance. If you have a critical safety issue, then it should probably take precedence and may not impact any of the other improvement focus areas. But, in general, because you will be expending time and money for process improvement, you would want to impact multiple focus areas.

Hoshin Action Plan

Next, you need to develop your Hoshin Action Plan. This further drives down your core objectives into the daily activities of your organization for process improvement by creating a detailed action plan. You should present this action plan to your leadership at a set frequency (typically a weekly walk through or monthly management review).

Figure 3.7 illustrates the Hoshin Action Plan. The top portion of the Hoshin Action Plan provides the necessary information to show the linkage between your action plan and each of your strategic core objectives. The following information is necessary:

- Core objective
- Management owner
- Department
- Team
- Date
- Next review

<table>
<tr><th colspan="3">Hoshin Action Plan</th></tr>
<tr><td colspan="2">Core objective:
Management owner:
Department:</td><td>Team:
Date:
Next review:</td></tr>
<tr><td colspan="3">Situation summary:</td></tr>
<tr><td colspan="3">Objective:</td></tr>
<tr><td>Short-term goal:
Long-term goal:</td><td>Strategy:</td><td>Targets and milestones:</td></tr>
</table>

FIGURE 3.7
Hoshin action plan.

To illustrate the Hoshin Action Plan, let's continue on with the example of improving product quality.

The next section is the situation summary. The situation summary provides a problem statement of the current status. It should clearly state why the improvement is necessary. An appropriate situation summary might be:

> Product quality is a key market driver in our industry. The external PPM for Product A has increased from 3,861 PPM to 4,725 over the past six months. As a result, our supplier rating as dropped from an A in the first quarter to a B in the second quarter.

Next, define your overall objective. This objective should relate back to one of your core objectives. The objective statement might then be:

> to improve product quality by decreasing external defects by 50%.

The next step is to complete the short-term and long-term goals using the metrics you previously detailed in your Hoshin Strategic Plan Summary (Figure 3.5). You should identify your short-term goals for a period of the next three to six months, and your long-term goal should focus on improvements for the next 12 months.

The next step is to discuss the implementation strategy. This should flow down from your Hoshin Plan Summary (Figure 3.6). At this point, however, it should be more detailed. For example, with the core objective of "improving product quality," your implementation strategy in the Hoshin Plan Summary might have simply stated, "Six Sigma Project." In the Hoshin Action Plan, you would want to clarify this in more detail: for example, you might explain this as: "Six Sigma Project on oversize cylinder bore."

The final step in the Hoshin Action Plan is to outline the targets and milestones of your strategy. In continuing with the example of the Six Sigma project on the oversize cylinder bore, your targets might be "perform a hypothesis test" or "run a design of experiments." The milestone would be the anticipated completion date.

Hoshin Implementation Plan

Now you should develop your Hoshin Implementation Plan, which records your progress and lists the implementation activities. Figure 3.8 is a template you can use for this. Your implementation plan compares the current status of milestones to your initial projections. It is typically shown in a Gantt chart format.

Hoshin Implementation Plan

Core objective:

Management owner:

Date:

Strategy	Performance		Schedule and Milestones Jan	Feb	Mar	Apr	May	June	July	Aug	Sept	Oct	Nov	Dec
	Target													
	Actual													
	Target													
	Actual													
	Target													
	Actual													

FIGURE 3.8
Hoshin implementation plan.

Review your Hoshin Implementation Plan with your organization's senior leadership at a set frequency, typically monthly. This requires each department to outline its expected improvement gains by month for the following year.

The top portion of the Hoshin Implementation Plan details each core objective, its management owner, and the date you're targeting to achieve that objective. Your core objectives on the Hoshin Implementation Plan should link back to the high-level Hoshin Strategy Summary Plan (Figure 3.5), and the management owner should link back to the Hoshin Plan Summary (Figure 3.6).

In the first column on the left of Figure 3.8, list your implementation strategies, as outlined by each department. Each department should have a Hoshin Action Plan for each implementation strategy.

In the next column of Figure 3.8, define your target and the actual performance for each implementation strategy. The performance should be measured using the metrics defined in your Hoshin Strategic Plan Summary (Figure 3.6). Then, break down the performance by month to monitor the performance improvement trends for the year. One way to visually show which metrics are on track by month is to color the background of that month's performance in green; months that did not meet the target performance can then be colored in red. This makes it easy for you to hone in on those implementation strategies that are not meeting their target performance.

Hoshin Implementation Review

Finally, you should conduct a Hoshin Implementation Review, which records the progress of your performance. Figure 3.9 provides a blank template. The Hoshin Implementation Review also records your company's performance relative to your industry's overall performance. The implementation plan also lists your highest priority implementation issues.

During your presentation to your senior leadership team, use your Hoshin Action Plans as backup information to show what targets and milestones you have met as well as a recovery plan to get performance back on track.

There are five key steps for implementing an effective strategy, described in the next sections.

Step 1: Measure your organization's system performance

In measuring your organization's system performance, it is critical to develop a plan to manage the strategic change objectives. The initial direction must be adaptable. The planning process must also be adaptive to respond to business changes.

Hoshin Implementation Review	
Core objective: Strategy owner: Date:	
Performance status	Implementation issues

FIGURE 3.9
Hoshin implementation review.

Then, regular assessments of planning and implementation are necessary.

Step 2: Set your core business objectives

To set your core business objectives, a technique called "catchball" is effective to incorporate group dialogue. Catchball is equivalent to tossing an idea around which provides the optimal objectives for the overall business system.

Step 3: Evaluate your business environment

The business environment must then be evaluated to understand the needs of the organization's customers. These customers include stockholders, employees, external customers, etc. The environmental analysis includes the technical, economic, social, and political aspects of the business. The purpose is to answer the question – How does the business perform relative to their competitors?

Step 4: Provide the necessary resources

For the strategic alignment to be successful, management must also provide the necessary resources to lead the efforts for both the strategic objectives and daily management. Remember – the purpose of Hoshin is to align the system to strategic change initiatives. This requires resource commitment.

Step 5: Define your system processes

Another key aspect is to define the system processes. Hoshin enables consensus planning and execution between all levels of the organization, as shown in Figure 3.10. The Hoshin Plan aligns the strategic vision, strategy, and actions of the organization. The actions of senior management, middle management, and the implementation teams (all levels of the organization) are aligned around the common Hoshin Plan. The next section discusses the three main tools of Policy Deployment for gaining consensus.

The Three Main Tools of Policy Deployment

Policy Deployment is one of the pillars of Total Quality Management (TQM). TQM is based around Deming's Plan-Do-Check-Act (PDCA) cycle. The three main tools of Policy Deployment are as follows:

1. PDCA cycle
2. Cross-functional management
3. Catchball

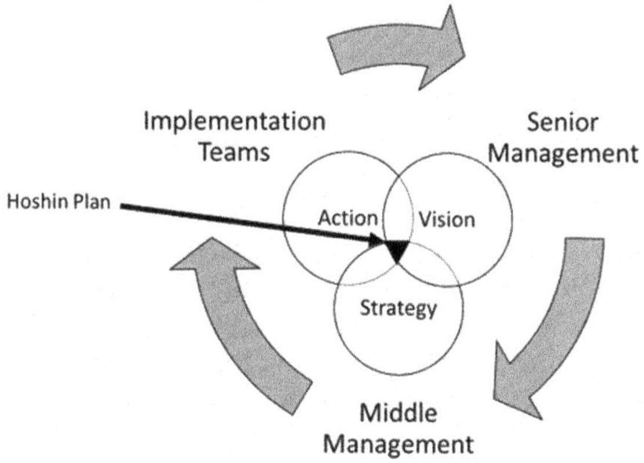

FIGURE 3.10
Hoshin plan alignment.

Let's look at each one in more detail to show how you can use these three tools for consensus planning and execution.

Deming's Plan-Do-Check-Act Cycle

Deming developed the Plan-Do-Check-Act cycle (shown in Figure 3.11) as an iterative four-step problem-solving process:

1. "Plan" consists of establishing objectives and processes to achieve specific results.

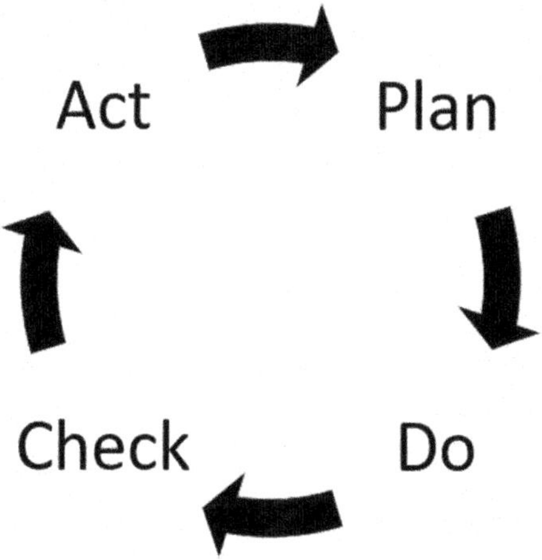

FIGURE 3.11
Deming's Plan-Do-Check-Act cycle.

2. The "Do" step involves implementing the processes.
3. In "Check," the processes are monitored and evaluated against the specifications.
4. In the fourth step, "Act," actions are taken to improve the outcome to meet or exceed the specifications.

One of the key differences between PDCA and Hoshin (aka Policy Deployment), though, is that Policy Deployment *begins* with the check step, which is Step 3 of Deming's cycle. Therefore, the cycle is really CAPD – check-act-plan-do. You start by checking the current status. This propels the Hoshin process. Each company-wide check begins the deployment of a new target and an action plan for achievement.

Cross-Functional Management

Cross-functional management (CFM) enables the continuous checking of targets and means throughout the product development and production processes. It is critical when developing the Hoshin Plan to include a cross-functional group from Marketing, Design Engineering, Quality, Manufacturing, Finance, Operations, and Sales depending on your organization's structure. This enables the diverse group to address the needs of all the shareholders (internal customers, external customer, stockholders, etc.) in the Hoshin Plan. By involving a cross-functional group, you can ensure a balanced representation of your customers' needs.

Catchball

Catchball involves continuous communication. It is essential for the development of targets and action plans. Catchball is also essential for the deployment throughout the organization. You must create feedback systems created to allow bottom-up, top-down, horizontal, and multi-directional communication. There must be commitment to total employee involvement.

Conclusion

Lean is a philosophy that must be fully embraced by an organization to truly reap the full potential. Unless you tie Lean to the strategic vision of your organization, your employees may view it as merely a "flavor-of-the-month" strategy – which, of course, means they will not embrace it, implement it, or even take it seriously. In addition, Lean projects that are selected based on their impact on the entire organization will have the most effective results.

You must ensure that your strategic vision cascades down throughout your organization into the daily activities of everyone. This clear linkage enables an organization to move in a common direction with common goals. Hoshin Kanri, also known as Policy Deployment, enables the strategic vision to be carried down throughout the organization.

When employees understand the direction of the organization, they can make the appropriate improvements that will enable long-term success. By using strategic vision, your organization can employ Lean techniques to eliminate waste and improve flow. As you target improvement opportunities, you will be able to see how they impact your organization's overall strategic vision. The alignment within your organization will also become clear.

4

The Tools

"The most dangerous kind of waste is the waste we do not recognize."

–Shigeo Shingo

If you think about the world, it is all based on continuous improvement. Every day and every aspect of our lives is based on it. We want to do continuous improvement on ourselves by being healthier or eating better. We want to make our work days easier by being more efficient. We continue to get the newest models of technology whether for our homes or cars or lifestyles. We must embed continuous improvement into our culture. In this type of culture, continuous improvement is a way of thinking and operating for people and the tools and techniques are simply a means to enable the business results and not the driver of activities.

Continuous improvement refers to all aspects of improving processes, products, methodologies, and techniques to ensure they provide more value to customers and are sustainable. This entails topics such as six sigma, lean, design for six sigma, quality, preventative and predictive maintenance and sustainability. Continuous improvement involves using various methodologies to design and measure changes to document the quantitative and qualitative improvements, and sustain the gains. This concerns breakthrough thinking and changing the culture of an organization. There are methodological and cultural aspects of driving continuous improvements in all types of organizations for sustainable results. The concepts behind continuous improvement will ensure companies innovate and become the best in class with successful financial results. Each company that does not embrace the operational excellence journey will lose their market to competitors.

In order to do this in our lean transformation, everybody in our organization solves problems and generates ideas to improve through the removal of wasteful activities as an integral part of our daily jobs. In a lean enterprise, we focus on reducing the wasteful activity, or non-value added time. By doing this we can have dramatic effects on the lead time and significantly improve our performance for the customer in terms of quality, service, reliability, and cost. With added flexibility, we can provide our customer with the maximum value that they expect.

DOI: 10.1201/9780429184123-4

The biggest part of continuous improvement we need to understand is that we need to work smarter and not harder.

Continuous improvement involves structural standardization where the effects build on each other. We need to understand what the standard is and then deviate from that standard. Returning to the standard is not continuous improvement but maintaining the current standard. Continuous improvement will now create an improved new standard.

There are different kinds of improvements.

People have a wealth of knowledge and talent that can help to improve the operating procedures. Everyone uses their problem-solving skills and tools and not only a few "experts." We must make time to make the small, everyday improvements on minor issues that trouble us daily. This will provide a structure to help manage all improvement ideas and enable easy knowledge sharing. A philosophy of continuous improvement is based on setting standards aimed at eliminating waste through participation of all employees.

Continuous improvement aims to eliminate all wastes in the operation by involving and utilizing skills and abilities of the company's workforce by focused continuous improvement activities. We must also focus on the people by revitalizing the culture and making the working environment fun, safe, and enjoyable.

In Lean thinking we use Kaizen on a daily basis, driving continuous improvement through the relentless pursuit of excellence and it is never ending, with everything that we do today being considered as the starting point for an improvement.

Kaizens are small, quick improvements to the current process, generally not requiring an event or a big team to arrive at the solution. Typically we can make the improvements within a month. We also give the authority and empowerment to the employees in processes that affect them daily. Kaizens also lead us to be able to make mistakes and try again which is the main principle of continuous improvement.

To institutionalize continuous improvement we need to ensure the robustness of all our processes. The robustness is achieved by standardized work, management control and accountability, leader standard work and visual cues that show us when we are not doing what we said we were going to do.

Leader standard work is a visual daily performance management system that is

1. Visual,
2. Timely, and
3. Drives both action and learning.

Leader standard work (LSW) is often the "missing link" to sustaining improvement.

Leader standard work ensures all the standard work are working as a system as intended and institutionalizing all processes.

Leader standard work focuses on accountability. It is a powerful tool to help leaders shift behavior to focus on the processes. It also challenges leaders to become teachers as well as problem solvers while separating the people who are unwilling to participate.

In order to sustain continuous improvement, top management must lead by example, going to the workplace which is what we call the "Gemba." We must look at the day-to-day, hour-by-hour problems and challenges. This will not give people the answers but the time, follow-up, and support to solving them.

Companies blame poor management or corporate when diminishing profits occur, but the root cause is from a lack of a lean system incorporating visual communication. Information not being portrayed is the biggest cause of problems because the communication aspect was lost. The continuous improvement aspects of increasing operational excellence while communicating the gains will make a successful workplace. The improvement strategy used is what makes a company successful. Not only should a company have these measures in place, but they need to ensure the sustainability of old improvements are taking place while incorporating new and continuous improvements daily.

Visual communication requires self-discipline and self-motivation. It takes an incorporated group for the fundamentals to work and cannot be successful with just one person. This initiative needs top management and must be driven to meet customer satisfaction. Visual standards and controls are a means of communication for all work aspects. They tell you when something needs to be done, where someone needs to be and when, how the progression of a project is going, and many safety measures.

It is important to know what to do with the communication and information given. Data is useless unless performance is changed due to the data given. Visual communication is the means to take data and make meaningful progressions to improve the workplace with sustainable performance metrics. Companies must grow with realistic expectations and then must be the best in the business in order not to lose the business. Gaining alignment will allow customers to grow.

It needs to be remembered that information presented is not progress; therefore, the information given must lead to progress. Items need to be accomplished in order to improve rather than just communicating the information given. Knowing what information is present and what information to share is imperative for visual communication. Communicating too much unnecessary information will only lead to more chaos and confusion. Messages need to be relayed in an easy manner that is simplistic so that employees are not overwhelmed.

A lean enterprise is market driven; when customers cannot get something from one supplier or are not happy with that supplier, they simply go elsewhere to a competitor. This is the importance of continuous improvement. There is always room for improvement no matter what we do. We must eliminate the variation in a system so that we have a standard. Once we meet that standard, we continuously improve upon it and raise that standard to the "Golden standard." Continuous improvement occurs in every step we take. Without continuous improvement, the world could not survive. The importance of continuous improvement should be taken very seriously because without it, we would live a life full of boredom and repetitiveness.

According to Edwards Deming, "It is not necessary to change. Survival is not mandatory." This quote shows the importance of continuous improvement being a key to our survival. However, we must remember, failure is only an opportunity for another round of continuous improvement, this time done with the data we have from our failures.

Change does not come easily when developing continuous improvement; however, the continuous struggle will make it inevitable that we need continuous improvement.

The concluding message is that without a concerted effort in continuous improvement, the economic system would not be able to survive along with ourselves. We always want to be better, so continuous improvement should be the key to life.

Main Tenants of Lean

It is important to remember that Lean is a choice. Nobody can make Lean happen. It is what you make of it. You can stay stagnant and keep doing things the way they've always been done. Or you can seek to continuously improve. The main concept of Lean is THE PEOPLE. Lean is not just about reducing waste and 5S, but about empowering the people, listening to the people. Understanding the people. Without the people, the business is not a business.

Continuous improvement refers to all aspects of improving processes, products, methodologies, and techniques to ensure they provide more value to customers and are sustainable.

The biggest part of continuous improvement we need to understand is that we need to work smarter and not harder.

Remember, the customer is the number one priority. Without the customer, we wouldn't be in business in the first place. So we have to respond to the new innovation and experiences the customer desires.

Then there's the process. We must work smarter and not harder!

Everybody in our organization solves problems and generates ideas to improve through the removal of wasteful activities as an integral part of our daily jobs. In a Lean enterprise, we focus on reducing the wasteful activity or non-value-added time. By doing this, we can have dramatic effects on the Lead Time and significantly improve our performance for the customer in terms of Quality, Service, Reliability, and Cost. With added flexibility, we can provide our customer with the maximum value that they expect.

We always want to be financially stronger so that we can grow our business. Once we grow our business enough, it is easy to use the financial backing to continue to grow our business.

And finally, it is all about the people. Empowering people is the way our business will grow in order to respond to the customer, reduce our waste, and grow the business. If we empower our people to want to be leaders, we will have more accountability, ownership, and opportunities (Figure 4.1).

This is the definition of Lean: The pursuit of perfection via a systematic approach to identifying and eliminating waste through continuous improvement of the value stream, enabling the product or information to flow at a rate determined by the pull of the customer.

Remember, we are trying to eliminate waste by continually improving using Lean tools to satisfy the customer.

Lean is NOT the next headcount reduction exercise, but Lean creates opportunities for doing more value-added activities. Lean will not succeed if the initiative stays limited to operations. Lean is NOT about working harder, but rather working smarter.

Traditional companies have some effectiveness that goes up and down over time. It is irregular, and there is no standardization or sustainability.

Lean companies strive to be the best by continually improving. They have standardization so there is sustainability and continuous improvement.

This is the satisfaction model for the customer. As quality goes up so does delivery while costs go down. We must fill three of these in order for the customer to be happy.

According to Toyota:

Quality is inherent in Toyota's products.

> Thanks to the company's constant striving for improvement (Kaizen), which has direct benefits for their customers. Toyota's insistence on maintaining quality throughout the production process is vital to ensuring that their finished products are of the highest quality.

Cost is always an issue.

> By buying Toyota products, their customers can be sure of having made a good choice. Kaizen ensures that Toyota products feature the latest effective innovations, maximizing productivity. The quality of Toyota's products allows their customers to enjoy a high return on their investment.

Kind of improvement		Tools
Rapid problem solving	Small improvement to return to standard Short throughput time	3C form Fishbone TIMWOODS
Kaizen Small scale improvement Kaizen (event) Bigger scale improvement	Simple problem Improvement to/ determine the standard Can be implemented by one person Short throughput time More complex problems Involving dedicated support and relevant people from specific process steps Medium throughput time	5x why Kaizen form A3 form One Point Lesson 8 steps of problem solving Impact/Effort
Project Big & complex improvement	High complex problems Many stakeholders involved Longer throughput time is expected (3 – 6 months)	Project Management

FIGURE 4.1
Continuous improvement.

Delivery is right for each customer's order.

Toyota's customer-driven system ensures that production output corresponds with timely delivery. Toyota's smooth, continuous, and optimized workflows, with carefully planned and measured work-cycle times and on-demand movement of goods, allow them to consistently meet their customer's expectations (Figure 4.2).

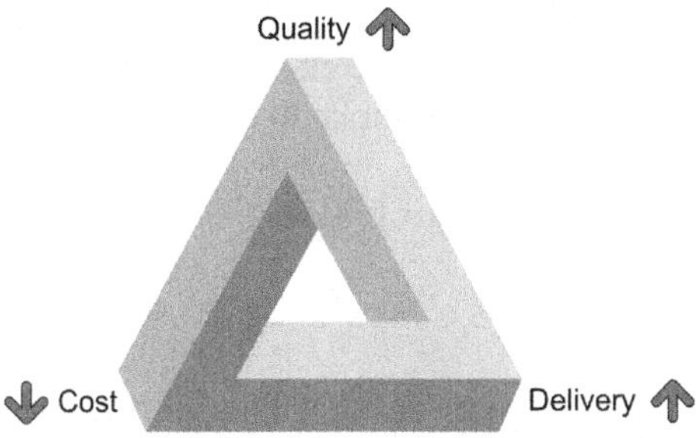

FIGURE 4.2
Triangle model.

There are a few key concepts of Lean.

We need to understand the customer and what exactly it is that they want. Then we need to have a continuous flow of production where we aren't interrupting our process with waste. We must pull materials showing when the start and finish of a product is. Each process pulls for the next. We must always eliminate waste. Waste is any non-value-added activity (Figure 4.3).

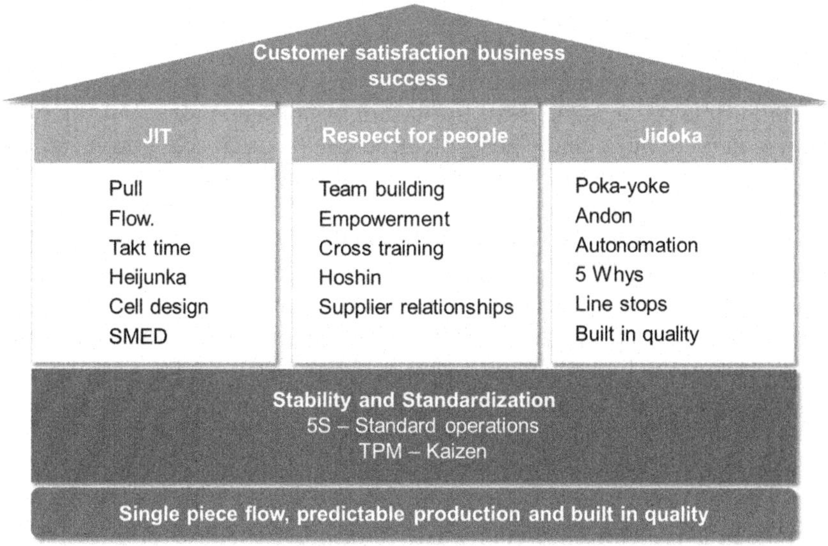

FIGURE 4.3
House of Lean.

A typical house of Lean has the following:

It starts with the foundation – 5S, cleanliness, Kaizen to continuously improve, total productive maintenance to ensure we have quality maintenance processes.

Then there are the pillars of our house.

Just-in-Time

Pull – There are three basic types of pull system; replenishment pull, sequential pull, and mixed pull system with elements of the previous two combined (see glossary at end). In all three cases, the important technical elements for systems to succeed are as follows: (1) flowing product in small batches (approaching one piece flow where possible); (2) pacing the processes to takt time (to stop overproduction); (3) signaling replenishment via a kanban signal; (4) leveling of product mix and quantity over time.

Flow – A continuous flow process is a method of manufacturing that aims to move a single unit in each step of a process, rather than treating units as batches for each step.

Takt time – Takt time is the rate at which work must be performed for customer demand to be met on time.

Heijunka is leveling, and this is done in order to meet demand while reducing waste.

A lean cell design ensures waste is minimized and the process flows using a pull system.

Single Minute Exchange of Dies (SMED) will minimize changeovers.

The middle pillar is the most important because it is all about the people. We need to build teams, empower people through cross training. We must understand the management vision through Hoshin Planning and finally understand the supplier and have a good relationship with them.

Finally, there is Jidoka, which is the quality at the source. Poka-yokes eliminate defects from happening in the first place. Andons are signals to signal a defect or problem. Autonomation empowers employees to build in quality every day. We must ask five why's to understand the root cause. We must stop the line when we have a problem so we can fix it right away. Built in quality will help us finalize the pillar.

All of this leads to a happy customer and a successful business.

- Lean systematically aligns people and processes with our strategy.
- Lean is the elimination of waste to improve the flow of information and material.
- What happens when you don't eliminate waste?

- It adds cost to the product/service with no corresponding benefit.
- It destroys competitive advantage.
- It makes work frustrating.
- It uses valuable resources (i.e., your time) to produce no value.

Working in a wasteful environment is frustrating. There is no forward-looking path or strategy to standardize work or continuously improve processes. There is a lot of variation in work because there is no set way to do things.

Value added versus non-value added

- Value-added time is anything the customer is willing to pay for.
- Non-value-added time is anything that does not add form, feature or function which the customer does not want to pay for (Figure 4.4).
- There are three main types of waste: Mura, Muri, and Muda.
- Mura is unevenness.
- Muri is overburden.
- Muda is pure waste.
- Mura can be solved with proper forecasting techniques.
- Muri can be solved with proper line balancing techniques.
- Wastes should be looked for through all processes and ultimately removed.

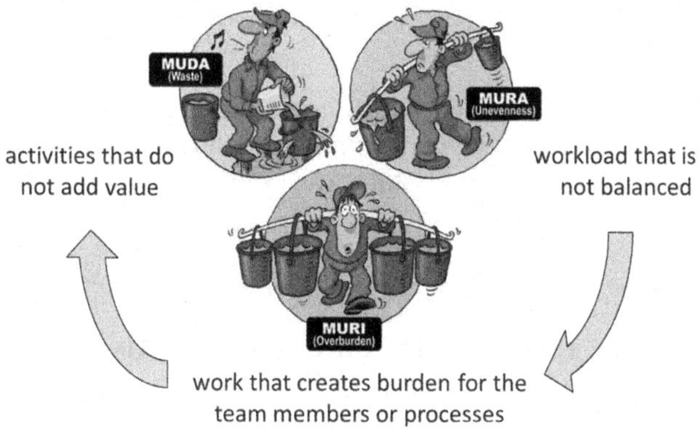

FIGURE 4.4
Mura, Muri, and Muda.

Muda is pure waste. We do not want any of this type of waste in our processes because it is costing us valuable time and money. This also does not add any value to our process or products. The customer does not want to pay for waste.

Muda

- A Japanese term for anything that is wasteful and doesn't add value.
- Waste reduction is an effective way to increase profitability.
- Waste occurs when more resources are consumed than are necessary to produce the goods or provide the service.
  - Anything that doesn't add value to the process.
  - Anything that doesn't help create conformance to your customer's specifications.
  - Anything your customer would be unwilling to pay you to do.

Here are examples of the eight types of wastes.

Defects – Efforts caused by rework, scrap, and incorrect information

Overproduction – Production that is more than needed

Waiting – Wasted time waiting for the next step in a process

Non-utilized talent – Underutilizing people's talents, skills, and knowledge

Transportation – Unnecessary movements of products and materials

Inventory – Excess products and materials not being processed

Motion – Unnecessary movements by people (e.g., excessive walking)

Excessive processing – More work or higher quality than is required by the customer

A Lean enterprise will improve quality, reduce lead time, and reduce costs.

The main process for a Lean enterprise is to map the value stream, identify the waste, create the future state, and continuously improve.

Safety, Quality, Delivery, Inventory, Productivity Metrics

Tier 1 Visual Boards

1. Boards are a visual management tool providing an almost real-time representation of area performance minimizing

cross-functional interfaces and drive rapid problem-solving and continuous process improvement.

2. Boards should contain three components: performance metrics, people, and continuous improvement (CI).
3. All boards shall have a list of Team Leads and meeting times posted (Figure 4.5).

Performance Metrics

1. Safety, Quality, Delivery, Inventory, and Productivity (SQDIP) indicators shall be present in every Tier 1 board.
2. Board metrics should be linked to site/business success metrics. Not every metric in the scorecard has to be in Tier 1 boards, use the 80–20 rule.
3. Metrics should be introduced step by step to facilitate Team Leader and team members' understanding. Involve employees during board design but be firm in terms of maintaining a set of metrics that are relevant to the operation. You can compromise on the format but be stricter in terms of content.

Observation Intervals

1. Boards must capture today's performance and trend over any meaningful interval, depending on the process. Determine observation intervals according to the process.
2. While fast moving processes may require hourly observations, slow pace processes may work well with shift observation intervals.

Trigger Points

1. Identify performance trigger points for corrective action for each metric. All metrics shall be actionable at the Tier 1 level. Follow the convention: Green = Good; Red = Action required.
2. Once a trigger point is exceeded, immediate containment action and rapid response problem-solving must be initiated, followed by appropriate countermeasures to avoid recurrence.
3. A process shall be in place to escalate issues beyond the team capabilities for resolution at a higher tier, involving support and other resources as appropriate (Figure 4.6).

Performance – Safety

1. The target for safety shall always be zero injuries.

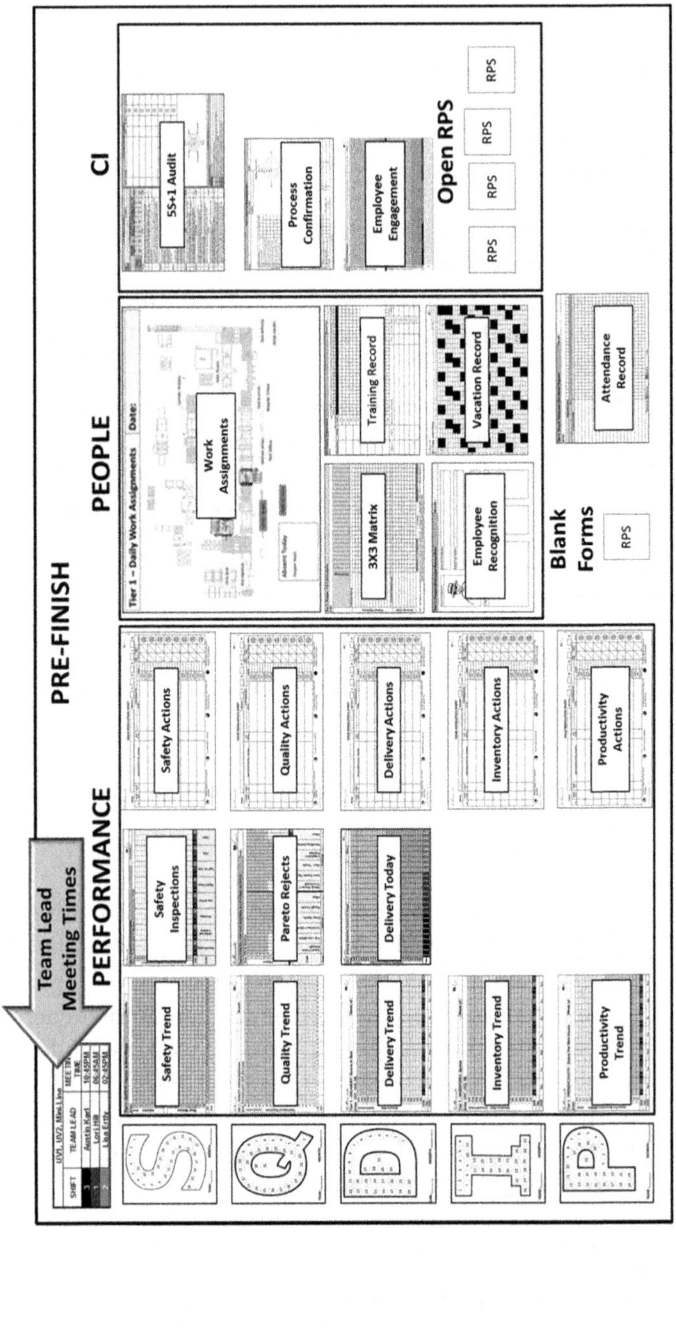

FIGURE 4.5
Tier 1 visual boards.

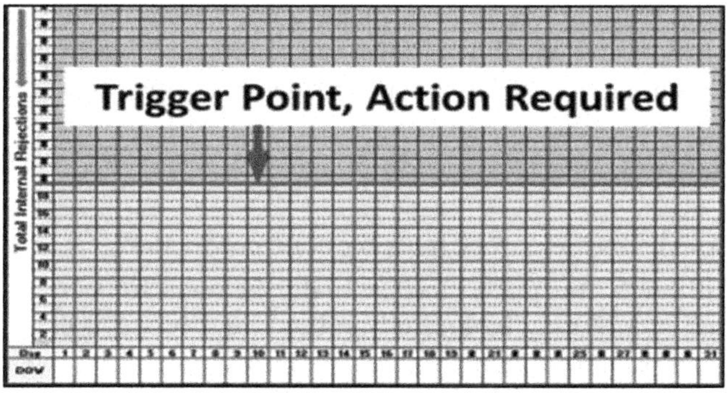

FIGURE 4.6
Trigger points.

2. In case of an injury, the first priority must always be providing timely care to the injured, according to Environmental Health and Safety (EH&S) plant emergency response procedures.
3. Escalate incident/injury according to site procedures.
4. Any EH&S incident shall be investigated and preventive and corrective actions put in place to avoid recurrence.
5. Any near miss and any hazard with potential to produce an injury must be treated with the same diligence as an injury.

Note: Start with the required forms and introduce other forms as the team matures on the basics and can manage other more complex forms. Depending on the nature of the operation, sites may choose to also track environmental issues where applicable. Metrics are process specific and will not be possible to use the same set of metrics in all deployments. All forms provided under visual are for reference only.

Performance – Quality

1. Trigger points for quality may be defined based on historical data as an aggregate number or by type of defects.
2. Set trigger points using the 80–20 rule, statistical process control (SPC) chart rules, or any other applicable criteria.
3. Defects may be classified as internal to the area and external (supplier process quality). Internal defects are mostly solvable in the area, but external defects almost always require escalation.
4. Where there are multiple product families, track quality by like products separately if it does not make sense to combine in a single chart.

5. A rapid response process to major quality issues and an escalation process for the flow of information must be put in place. Site management has to define what a "major" quality issue is.

Performance – Delivery

1. Use pitch charts to track hour-by-hour production.

   Although the pitch chart actually belongs to Tier 0, in one stream area it may be located on the Tier 1 board if close to the production line. When an area has multiple streams, it is recommended to place pitch charts close to the line and post summary data on board charts.
2. At Tier 0/Tier 1, delivery must be production counts.
3. When products have different cycle times, priority boards or completion Heijunka may be used to track deviations from standard processing time.
4. Always post a schedule on the board with actual shop orders to be delivered today. It is important to produce the right quantity but it is equally – if not more– important to produce the right products.

Performance – Inventory

1. To maintain flow, a certain level of inventory is required; however, too much inventory is waste and too little inventory may produce flow interruptions.
2. In a pull system, inventory always belongs to the producer process.
3. When inventory levels are controlled via Kanban, the inventory metric may be percent of Kanban within limits.
4. Other inventory metrics may apply depending on the operation.

Performance – Productivity

1. Productivity is a measure of how efficiently the plant uses its labor.
2. Generic formula:

$$\text{Productivity} = \frac{(\text{Planned Staffing} \times \text{Actual Production})}{(\text{Actual Staffing} \times \text{Planned Production})}$$

3. Practical formula:

$$\text{Productivity} = \left(\left(\sum_{i=1}^{n} \text{FTT}_i\right) \div \left(\sum_{i=1}^{n} \text{PD}_i \times \text{HC}_i\right)\right) \Big/ (\text{Std PPMH})$$

where FTT is the first time through, PD is the pitch duration, HC is the headcount, all in pitch "i," and PPMH is parts per man-hour.

4. These formulas are ratios that can be expressed as percent. Assuming an hourly pitch, an alternate formula that will provide the same information in a different scale is:

$$\text{Actual PPMH} = \left(\sum_{i=1}^{n} \text{FTT}_i\right) \Big/ \left(\sum_{i=1}^{n} \text{HC}_i\right)$$

All these values should be available on the pitch chart.

Issue Resolution

1. Issue Resolution Forms should be associated with each of the SQDIP metrics. Floor issues and metrics beyond trigger points must be logged on those forms.
2. Containment actions and countermeasures shall be identified, assigned for resolution, and tracked through completion on a daily basis.
3. Long-term actions that cannot be resolved during the same day should have a rootcause problem solving (RPS) form and actions split into smaller daily tasks.
4. Actions for today must be prioritized and signaled with a priority start or any other means.
5. The Plan Do Check Act (PDCA) circle on the forms shall be updated accordingly to instruction on footer of forms.
6. An escalation process should be in place for actions the team cannot resolve, requiring additional resources to be involved.
7. Escalated actions may be identified using an Escalation Sign and/or an Escalation Form (Figure 4.7).

People – Employee Attendance

1. An attendance sheet shall be used at each board and each employee must check presence at the tier meeting.

Suggestion & Issue Resolution									
You					Owner				
ID Number	SQDIP Metric / 5S	Posted Date:	Initiated By:	Improvement / Concern / Issue	Actions	Owner -Or- Escalated To:	PDCA Status	Est. Due Date	Date Closed
Ex.	Q	12/1	Tina A	Job Aid missing for Line 1	Create Standard Work for Line 1	T2	A P C D	12/10	12/10
1							A P C D		
2							A P C D		
3							A P C D		
4							A P C D		
5							A P C D		
6							A P C D		
7							A P C D		
8							A P C D		
9							A P C D		
10							A P C D		
11							A P C D		
12							A P C D		
13							A P C D		
14							A P C D		
15							A P C D		
P - Plan: Define the problem needing to be addressed, D- Do: Develop and implement a solution, C- Check: Confirm results of solution, A- Act: Document the results									

FIGURE 4.7
Issue resolution log.

2. Team Leads use this information to populate the Daily Work Assignments (see Task 12).
3. Optionally, a weekly Absence Control Form by day and shift may be used across the board or for problem areas only.
4. Plant management should define an acceptable level for a Tier 1 area. It is recommended to set the trigger point of two or less.

People – Daily Work Assignments

1. Use an area layout or any other means to identify work stations in the area.
2. During the tier meeting, Team Leads place preprinted name tags or write employee names at each workstation using the Meeting Attendance Sheet.
3. To cover for vacations or unplanned absenteeism, the Team Lead will use the 3 × 3 matrix.

People – Overtime Tracker

1. Overtime affects plant productivity. The overtime tracker contains area employee names and columns with planned and actual overtime hours for each day of the week. Other site existing formats may be equally used if available.
2. This form has the dual purpose of helping with productivity calculations and ensuring all employees in the area have an equal opportunity to work overtime.

People – Vacation Planner

1. The Vacation Planner is a form containing days of the month across columns and months across rows with weekends and holidays identified for current year.
2. Each month has three rows for three names at a time maximum. Other site existing site formats may be equally used if available.

People – 3 × 3 Skill Matrix

1. The skill matrix contains a list of employee across rows and jobs or workstations across columns.
2. Each job must have a minimum of three qualified operators and each operator must be proficient in at least three jobs (3 × 3).
3. Color code cells green for "Qualified," yellow for "In training," and red for "Training Priority."
4. Information from the skills matrix is used to populate the training plan.

People – Training Record

1. Once training needs are identified in the 3 × 3 skills matrix (see Task 15), training must be scheduled in the Training Record Form.
2. Training completed in the planned date receives a green dot. Past due/late training receives a red dot.
3. The reason why training was not conducted as planned must be investigated, and a corrective action should be initiated, as described under Issue Resolution Form.

People – Employee Recognition

1. The site should have an employee recognition program based on contributions to site continuous improvement efforts.

2. Employee accomplishments should be documented, posted on the board, and communicated to the team during regular tier meetings.

CI – Change Management

1. A form should be posted on the board and discussed during regular tier meetings as appropriate to communicate changes affecting the area such as new or updated standard operating procedures (SOPs), new materials or processes, etc. This communication comes normally from upper tier or functional leaders.

CI – 5S + 1 Audit and Inspection Results

1. 5s+1 is a basic visual form of Lean management. Periodical 5s+1 self-inspections and audits will occur and the area must address findings from these observations, and use results – generally plotted in a form of a radar chart – to drive improvement projects.
2. Corrective action should be initiated for each audit finding/opportunity for improvement, as described under the Issue Resolution Form.

CI – Process Confirmation

1. Tier 2 or higher Team Leaders or Team Leads from other areas will conduct documented observations during tier meetings, using the Process Confirmation Sheet.
2. Observations must be shared with the team in the form of coaching and results posted for team analysis and to drive improvements (Figure 4.8).

CI – Employee Engagement

1. Employee continuous improvement involvement should be documented and tracked in the form of employee engagement.
2. Each plant may define what engagement indicators to use.

   As an example consider number of number of CI projects raised and CI projects closed by employee, number of Kaizen events employee participated, etc.

Process Confirmation Assessment

Team Leader: **Date:** **Coach's Name:**

Team Meeting

	Met	Not Met
1. Meeting Agenda		
a. Attendance is verified at the beginning of the Tier meeting. All personnel are present and on time.		
b. Performance metrics and CI are covered during the Tier meeting by the action owners		
c. Thank yous, recognitions, coaching are conducted.		
d. Team Leaders carry LSW and there is evidence of task completion.		
2. Information Board Review		
a. Information and charts are updated with current information and goal/trigger points.		
b. Board has updated pictures and last version of all standard forms. Board is clean and in good shape.		
3. Running the Meeting		
a. Action is taken when trigger points are reached.		
b. Actions are assigned to a person in the meeting and recorded on the board.		
c. Action items are closed on time according to defined due dates.		
d. Action items effectively look to eliminate root causes of problems.		
e. There is traceability when action items are escalated.		
4. 5S + 1		
a. 5S+1 audit scores and graphs are updated.		
b. 5S+1 audit finding are addressed using the Tier escalation process.		
5. Continuous Improvement:		
a. There are blank forms of "Continuous Improvement Ideas" available.		
b. The team has proposed improvement ideas and they have been implemented.		
c. There is evidence that RPS opportunities are identified during the meeting.		
d. There is evidence of use and sign off of Change Management communications.		

Behaviors

	Met	Not Met
6. Team Leader - Meeting Presentation		
a. Meeting Space is adequate to accommodate the team all team members are close to team leader during meeting.		
b. There are two way communications, not just presenting board content.		
c. Team Leader ensures only one person speaks at a time and makes sure people's ideas are heard and respected.		
d. Speakers have clear, audible voice that participants are able to hear. They face the team while speaking making eye contact.		
7. Team Leader - Meeting Facilitation		
a. Team Leader encourages team for participation and provides guidance to prevent conflicts.		
b. Team members acknowledge assigned actions when assigned and provide preliminary assessment in 24 hours.		
c. Team Leader provides CLEAR TIMINGS and reminders to team. Facilitator provides guidance to team members on action resolution.		
d. Team Leader ensures problem solving does not occur during the meeting and encourages team to setup time to discuss problem resolution.		

Count # of "Met" and circle the number with the % to target for the meeting being audited

# of Met	%	*# of Met*	%
1	4%	16	64%
2	8%	17	68%
3	12%	18	72%
4	16%	19	76%
5	20%	20	80%
6	24%	21	84%
7	28%	22	88%
8	32%	23	92%
9	36%	24	96%
10	40%	25	100%
11	44%		
12	48%		
13	52%		
14	56%		
15	60%		

(1) Threshold Score Target > 90%

Please Note Improvement Opportunities Below:

FIGURE 4.8
Process confirmation example.

Visual Management Examples

These system status indicators are visual controls that can be understood at a glance (Figures 4.9–4.11).

Conclusion

Remember – it is important to monitor the results of your process that was improved. Understand the project goals and how you will measure

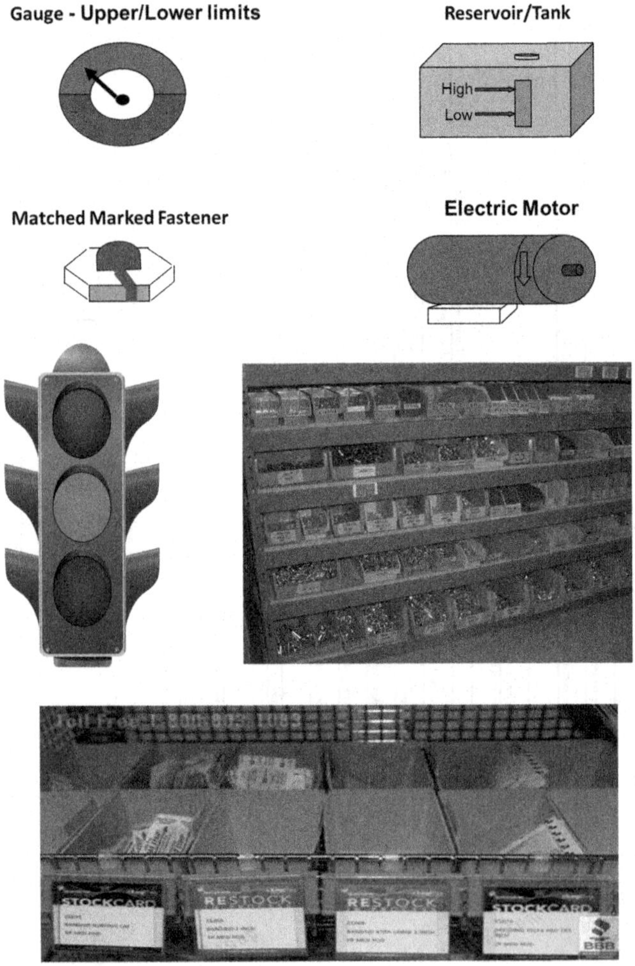

FIGURE 4.9
Visual management examples.

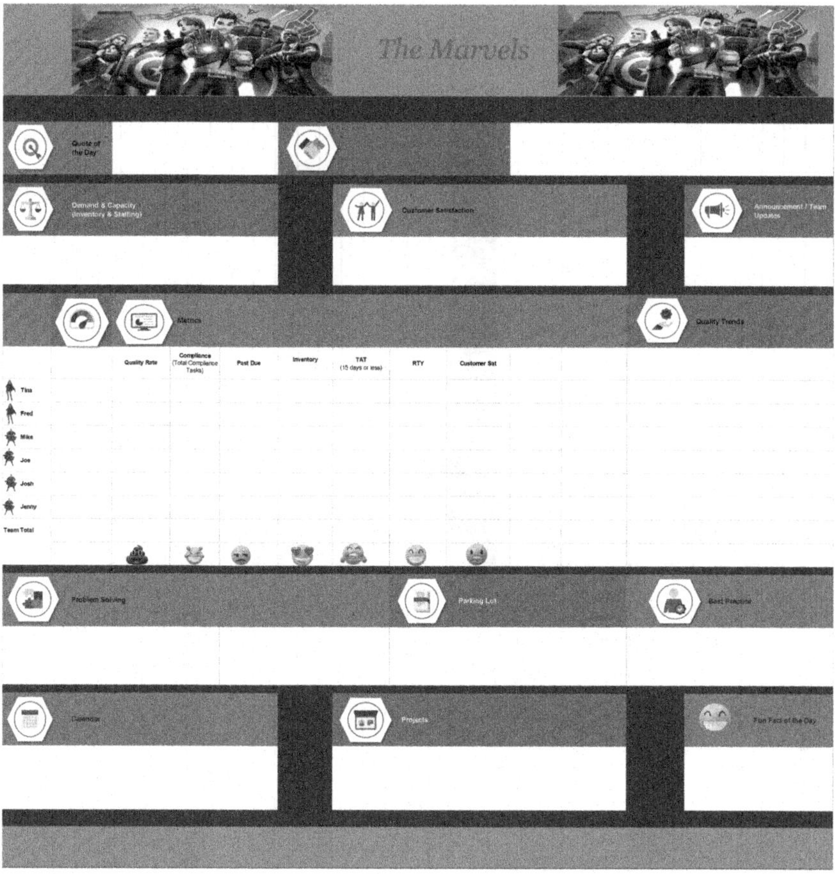

FIGURE 4.10
Virtual huddle board example.

the process. It will ensure success of your project. Understanding the data and processes with a view to specifications needed for meeting customer requirements should be sought after. Measurement systems should be developed and evaluated, and current process performance should be measured.

The first step is to identify the organization's key strategies and mission and translate the vision into operational goals for the organization. The executive team typically sets the strategies and mission based on the organizations key competencies and competition. The specific goals and objectives that need to be accomplished to support the strategic plan for the organization are then determined and communicated throughout the organization. This helps link the vision to individual

Day	Task	Date	Person Audited	Comments
M TU W TH F	Review T1 - T4 Boards			
M TU W TH F	Attend and coach a T1, T2, T3 or T4 Meeting per schedule			
M TU W TH F	Complete Process Confirmation			
M TU W TH F	Audit T4 Leader LSW			
M TU W TH F	Audit T3 Leader LSW			
M TU W TH F	Audit T2 Leader LSW			
M TU W TH F	Audit T1 Leader LSW			
M TU W TH F	Complete GEMBA Walk			
M TU W TH F	Meet with Plant Manager to debrief			
M F	Travel			
Weekly	Review 5S+1 Assessment			
Weekly	Review Target Tree T1 - T4 (verify with board alignment)			
Weekly	Review VSM and Kaizen activities			
Weekly	Speak with Operator in Lighthouse			
Weekly	Speak with Operator outside of Lighthouse			
Weekly	Speak with Tier Leader in Lighthouse			
Weekly	Speak with Tier Leader outside of Lighthouse			
Weekly	Speak with Quality Mgr/Engineer/Lead			
Weekly	Speak with Safety lead			
Weekly	Speak with BB, OD, PE that are on site.			
Weekly	Expense Report			
Weekly	Meet with Kris V			
Weekly				
Bi-Weekly	Meet with David M			
Bi-Weekly	Meet with Joan W			
Bi-Weekly	Meet with Carl			
Bi-Weekly				
Bi-Weekly				
Bi-Weekly				
Bi-Weekly				
Monthly				
Monthly	Complete any mandatory training			
Monthly				
Monthly				

Coaching Opportunities

	Date	Description	Initial
☐			
☐			
☐			
☐			
☐			

Thank-you's & Recognition

Who
For What?
Who
For What?
Who
For What?
Who
For What?
Who
For What?
Who
For What?
Who
For What?
Who
For What?
Who
For What?

FIGURE 4.11
Leader standard work.

performance. The metrics for each level of the company are then determined and targets set for business planning. These metrics should link directly to the organizational strategic goals and objectives. The final step is feedback and learning. Continuous improvement and communication are utilized to adjust the strategy as appropriate. A dashboard is utilized in order to view key metrics. It is important to understand the project goals from the beginning and how to measure the process. It should be understood what a leading and lagging indicator consists of. The project charter delivers the Y, or process outcome, by clearly stating what the business or process problem is. The Y is the lagging indicator or the final checkpoint at the end of the process. The previously identified Xs are called leading indicators. These are checkpoints during the process.

A performance measurement and monitoring system or baseline of Key Performance Indicators is established that helps to measure and control the critical leading Xs and lagging Y indicators continually. The Y should be a measurable process metric that tells how well the process is performing today, which is called the baseline.

The performance based on the baseline should be the improvements or the goals. Common examples of customer metrics include customer satisfaction rate, customer retention, referral rates, quality, on-time delivery, and number of new products launched this year.

Common learning and growth metrics include employee turnover, employee morale/satisfaction, percent of internal promotions, percent of succession plans completed, absenteeism, and number of employees trained as a Six Sigma Black Belt or Green Belt. Our internal processes must be effective at meeting the customer's requirements. And to keep costs low, they must be efficient. Finally, we will want the people to be capable, trained, and have what they need to operate the process in that manner. As you can see, tools are critical to driving your organization's performance and that performance should be seen through metrics.

Questions

1. What is the difference between value and non-value-added activities?
2. What are the eight main types of waste?
3. What are lagging indicators?
4. What is Muda?

5. What is the main principle of SMED?
6. What are the main components of flow?

Reference

Heller, R. (1999). *Managing change.* Dorling Kindersley Ltd.

5

Building a Sustainable Lean Culture Case Study: Process Improvement

- **Executive summary**: Argo is a growing product for a dog toy factory. The products are selling extremely well and sales have been rising for the past year. Over the past year, with the increase of sales, there has also been an increase in complaints for problems with the toys. Customers felt that some toys fall apart far too quickly. Poor material quality of the product results in major variation in product due to materials, methods, machinery, etc. Many toys were being reworked in the factory causing more time and money while some customers were complaining as well.
- **Problem statement**: Reduce rework for all toys by 15% by March 2021 and 25% by June 2021 and have less than 1% customer complaints.
- **Methodology**: Training, Equipment, Line-Balancing, Layout, Process, Data-Gathering Root Cause Analysis.

Introduction

- Tina Agustiady – Continuous Improvement Leader
- Buddy Lee – Operator
- Mike Thomas – Operator
- Brian Crops – Production Engineer
- Tamara Brown – Maintenance Coordinator
- Nelly Curtis – Plant Manager and Project Champion
- Leo Downs – Executive Sponsor
- Gweneth Verns – Master Black Belt
- The team was selected based on knowledge and expertise of the process and was proficient and organized during the project.

DOI: 10.1201/9780429184123-5

Define Stage

The processes committed for Argo are important to the customers and the business. A process map was completed in order to better fully understand the steps. The project team now has a baseline to begin the Measure phase through the process steps.

The following will include:

- Project charter
- Milestones and deliverables
- High-level process map

Project charter (Figure 5.1).
Milestone and deliverables (Figure 5.2).
High-level process map (Figure 5.3).

Conclusion of Define Phase

The Define stage showed that the processes committed for Argo are important to the customers and the business. A process map was completed in order to better fully understand the steps. The project team now has a baseline to begin the Measure phase through the process steps.

Measure Phase

The goals of the Measure phase were to determine the key factors for variation issues, machine issues, and variation in Argo processes through statistical analysis and graphical analysis. The goal of the project is to reduce rework by 15% by March 2021 and by 25% by June 2021 while having less than 1% customer complaints. During the Measure phase the following charts were created:

- SIPOC diagram
- Cause and effect (C&E) diagram (fishbone diagram)

Project Title: Argo re-Work Reduction

Black Belt	Project Champion	Executive Sponsor	MBB/Mentor
Margaret Sanchez	Nelly Curtis	Leo Downs	Tina Agustiady

Primary Metric	Secondary Metric
Toys Re-Work Cost. Measure materials and time wasted on fixing toys that are defective. Also include money for increase in labor to the factory to compensate flawed materials.	Reduce re-work for all toys by 15% by March 2021 and 25% by June 2021 and have less than 1% customer complaints.

Problem Statement	Business Case
Reduce re-work for all toys by 15% by March 2021 and 25% by June 2021 and have less than 1% customer complaints.	Goals: Decrease re-worked toys by 15% by March 2021 and an additional 10% decrease by June 2021. By June of 2021 we would like a total of 25% decrease of toys re-worked in the factory so that Argo can regain the trust of its customers and decrease the number of customer complaints to less than 1%. Achieving this will also allow Argo to have increase profits to meet or overpass financial goals for 2021.

High Level Project Timeline

Phase	Start	Finish
Define	1/1/2021	1/8/2021
Measure	1/11/2021	1/29/2021
Analyze	2/1/2021	2/24/2021
Improve	2/25/2021	2/26/2021
Control	3/1/2021	3/5/2021

Constraints & Dependencies	Project Risks	Other Diagnostics
Resources available: Knowledgeable employees who know that process of the product. Respectable budget to account for all expenses. Possible machinery issue.	Exceed deadline time allowed. Resources unavailable in timely fashion.	

Approval/Steering Committee		Stakeholders & Advisors		Project Team & SME's	
Name	Organization	Name	Organization	Name	Organization

FIGURE 5.1
Project charter.

Project Milestones	Due Date	Deliverable
Define Phase:	1/8/2021	Project Charter and Process Map
Measure Phase	1/29/2021	Data gathering, Cause and Effect Diagram, SIPOC, Cause and Effect Matrix,Failure Modes and Effects Analysis, FMEA, Capability Analysis, Guage R&R, Root Cause Analysis Setup
Analyze Phase	2/24/2021	Rooit Cause Analysis, Data and process Analysis
Improve Phase	2/26/2021	Implemenate Plan based on solutions
Control Phase	3/5/2021	Control Plan, Audit Checklist, and Project Conclusion
Final Report	3/5/2021	Project

FIGURE 5.2
Milestone and deliverables.

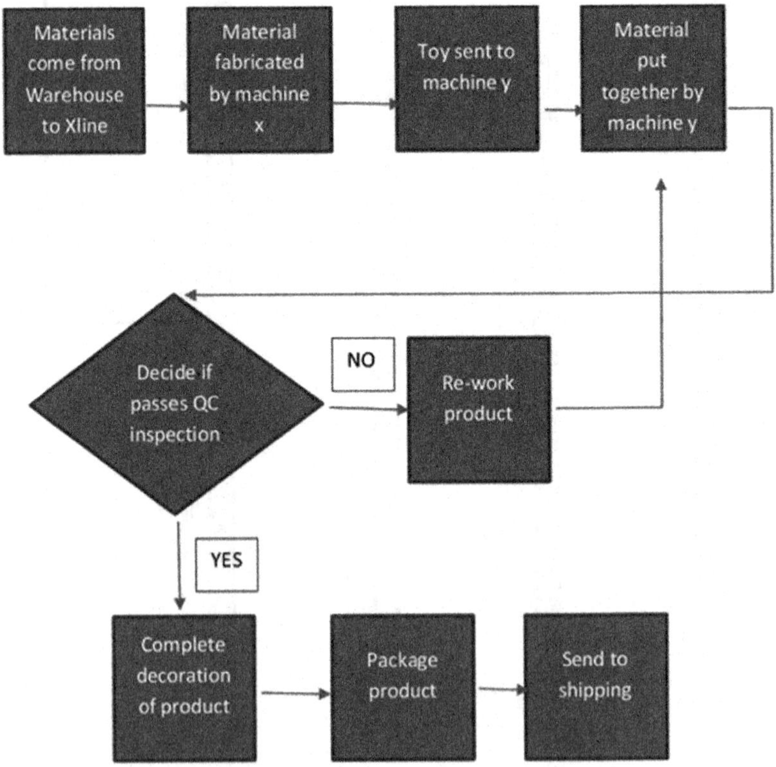

FIGURE 5.3
High-level process map.

- Cause and effect matrix
- Failure modes and effect analysis [a Pareto chart was completed for the failure modes and effect analysis (FMEA)]
- Pareto chart for the largest issues

SIPOC (Figure 5.4).

Cause and effect diagram (fishbone diagram) (Figure 5.5).

Cause and effect matrix (Figure 5.6).

Pareto chart for cause and effect matrix (Figure 5.7).

Failure modes and effect analysis (Figure 5.8).

Pareto chart for FMEA (Figure 5.9).

Pareto for customer complaints (Figure 5.10).

Conclusion of Measure Phase

- Data were taken of as many parameters as possible before changing any variables. It was found that material was making a significant impact on the process and there needed to be analysis on the FMEA and C&E matrix.

Suppliers	Inputs	Input Specification	Processes	Gap	Outputs	Customers
Raw Bone	Thread	1.0 - 1.1	Materials come from warehouse to Xl ine		Raw Material	Wally
Pet Doo	Material	2.0 - 2.9	Material fabricated by machine X		Fabricated Material	Tget
Cryer	Bone	4.1 - 4.9	Toy sent to machine Y		1/2 processed toy	Kroman
Whiner	Cushion	.8 - 1.0	Material put together by machine Y		Toy complete for inspection	Giant Store
Happy Pup	Squeaker	5.5 - 6.5	QC Process	**Manual Process**	Toy complete for inspection	Pet Peeps
	Rope	10.1 - 10.9	Re-work		Toy complete for inspection	
	Treat	2.4 - 2.10	Decorate product		1/2 processed toy	
			Package		Fully processed toy	
			Ship		Toy ready for customer	

FIGURE 5.4
Suppliers Inputs Processes Outputs Customers (SIPOC).

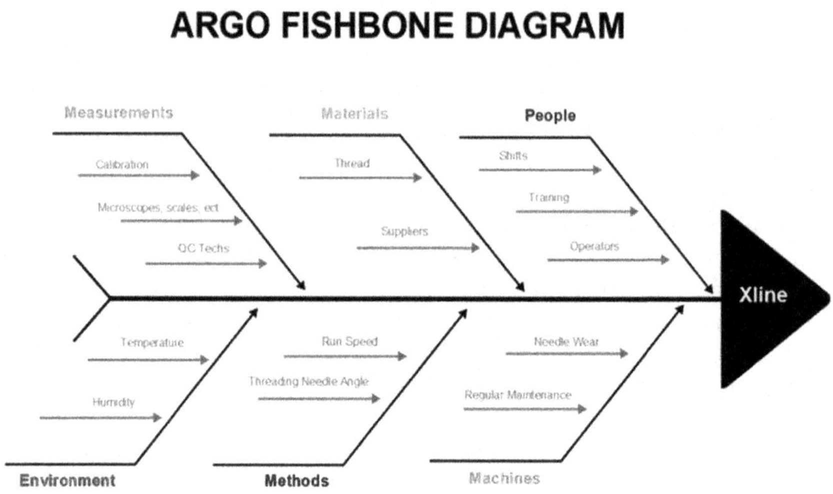

FIGURE 5.5
Cause and effect diagram (fishbone diagram).

Analyze Phase

- Since the material was automatically seen as in issue. The specifications were looked at. It was noted that there were two material suppliers:
  - Cryer
  - Whiner.
- The specifications were very different for the materials in terms of the thread count which made a big difference in the material quality.
- The material needed to be between 2.0 and 2.9 (thousands).
- Cryer's specifications had an average of 2.6 (thousands).
- Whiner's specifications had an average of 2.3 (thousands).
- It was noted that Whiner's means were shifted to the left and needed to be centered.

Mean shift analysis (Figure 5.11).

Box plot for Cryer versus Whiner (Figure 5.12).

Meaning of the box plot concludes:

- The box plot for Cryer versus Whiner indicates that the higher thread count average of Cryer has a higher mean than that of Whiner which

Cause and Effect Matrix

X Line Process		Rating of Importance to Customer	7	4	6	10	9	
Process Step			1	2	3	4	5	
			Toy's Design Parameters	Plush Filling	Preloaded Thread Colors	Completed Primary Sewing	Completed Detail Sewing	Total
	Process Step	Process Input						
1	Set Machine Parameters	Manual Key In	10	0	0	10	10	260
2	Select Plush Filling	J&J Plush Filling	5	10	0	0	0	75
3	Load Threading	Cryer or Whiner's Red 200	9	0	10	9	9	294
4	Load Threading	Cryer or Whiner's Blue 106	9	0	10	9	9	294
5	Load Threading	Cryer or Whiner's White 700	9	0	10	9	9	294
6	Load Threading	Cryer or Whiner's White 750	9	0	10	9	9	294
7	Load Threading	Cryer or Whiner's Black 8	9	0	10	9	9	294
8	Load Threading	Cryer or Whiner's Black 10	9	0	10	9	9	294
9	Computer Guided Sewing 1	Preloaded Threading	0	0	0	10	10	190
10	Computer Guided Sewing 2	Preloaded Threading	0	0	0	10	10	190
11	Loaded for transfer to packing	Toy Transfer Cart	0	0	0	0	0	0
Total			490	44	366	850	765	

FIGURE 5.6
Cause and effect matrix.

has a lower mean. Cryer average thread count is on the upper mean of the specifications of the material that is needed. Even though Whiner is on the lower end, the specifications are still within the specification range that is needed. Further data will be needed to determine the number of toy rejects because of the materials that are used.

Conclusion of the Analyze Phase

- According to the data between the two different materials, it can be one of the variables that can account for the rejected/reworked

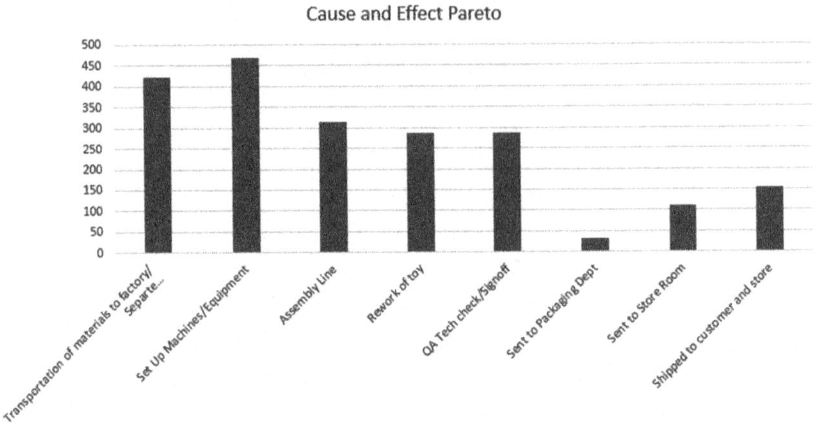

FIGURE 5.7
Pareto chart for cause and effect matrix.

toys. Other variables can be factors like staff, equipment, environment, and inspection of the materials. Further data will be needed to confirm that the quality of the materials have made an impact that will cause the amount of rework. They will also have to see when those materials were used and under what circumstances, so that they can create or at least mimic almost exact conditions to confirm that the materials are what caused the increase of rework.

Improve Phase

- The team decided to look at the Cryer supplier to ensure that they were giving good product. Graphical analysis was completed for Cryer (Figure 5.13).

Meaning of the graphical analysis for Cryer:

- The graph shows to follow the main principal of what a normal distribution curve looks like.
- Yes, it looks normal.

Hypothesis test (Figure 5.14).

The histogram of differences *t*-test (Figure 5.15).

#	Process Function (Step)	Potential Failure Modes (process defects)	Potential Failure Effects (KPOVs)	SEV	Class	Potential Causes of Failure (KPIVs)	OCC	Current Process Controls		DET	RPN
1	Materials come from warehouse to Xline	Bad material	Torn product	8	XX	Supplier giving bad materials	9	QA Check	8	8	576
2	Material fabricated by machine X	Bad material or bad fabrication	Torn product	8	XX	Machine x problems	7	QA Check	10	10	560
3	Toy sent to machine Y	Bad material or bad fabrication	Dismantled product	8	XX	Machine x problems	8	QA Check	8	8	512
4	Material put together by machine Y	Bad material or bad fabrication	Dismantled product	7	X	Machine y problems	8	QA Check	8	8	448
5	QC Process	Manual check not completed correctly	Bad product sent out	8	XX	QC not complete	7	QA Check	7	7	392
6	Re-work	Threading looks bad and customer complaint occurs	Bad product sent out	8	XX	QC not complete	7	QA Check	7	7	392
7	Decorate product	Decorating not properly completed	No or double decorating	8	XX	Machine does not notice already decorated	5	None	8	8	320
8	Package	Bad packaging	Bad product sent out	6	X	Packaging machine flaw	5	None	7	7	210
9	Ship	Shipping breaks product	Customer complaint for breakage	8	XX	Shipping flaw	6	None	2	2	96

FIGURE 5.8
Failure modes and effect analysis (FMEA).

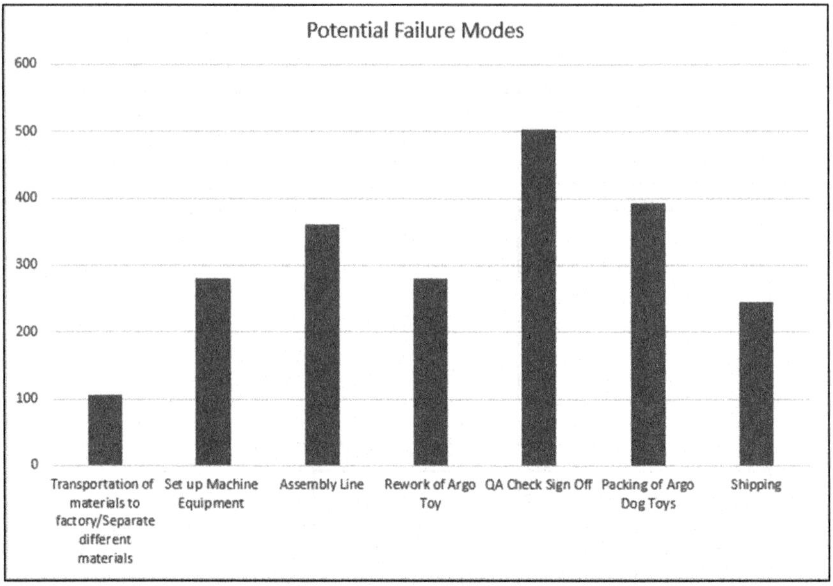

FIGURE 5.9
Pareto chart for FMEA.

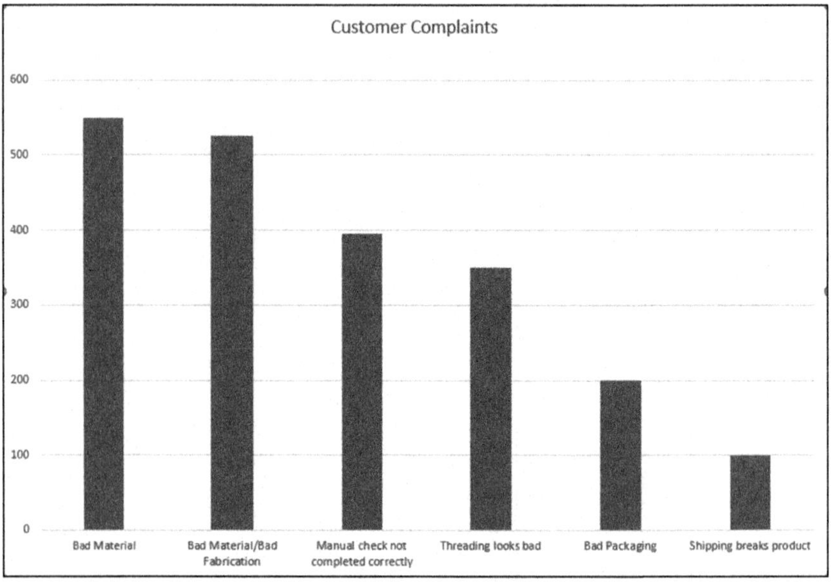

FIGURE 5.10
Pareto for customer complaints.

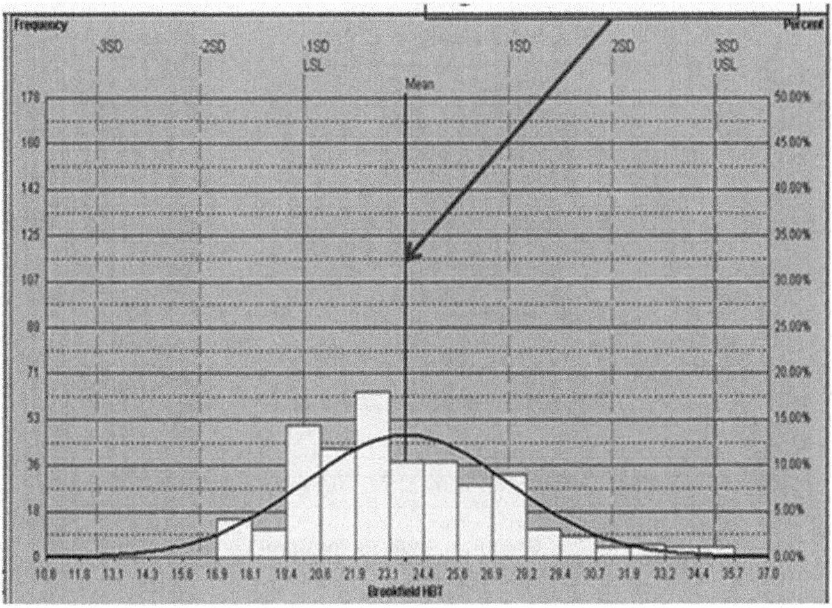

FIGURE 5.11
Mean shift analysis.

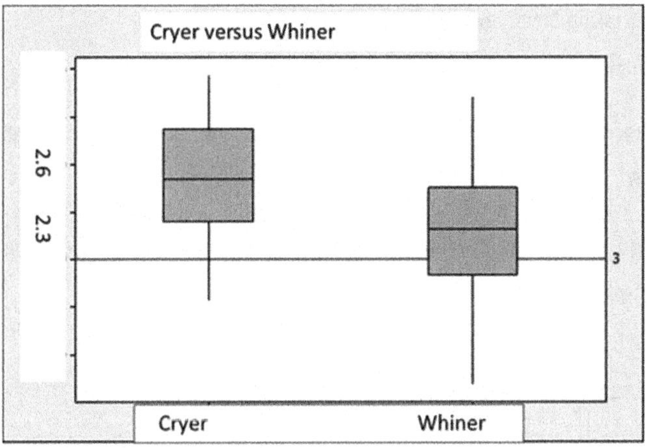

FIGURE 5.12
Box plot for Cryer versus Whiner.

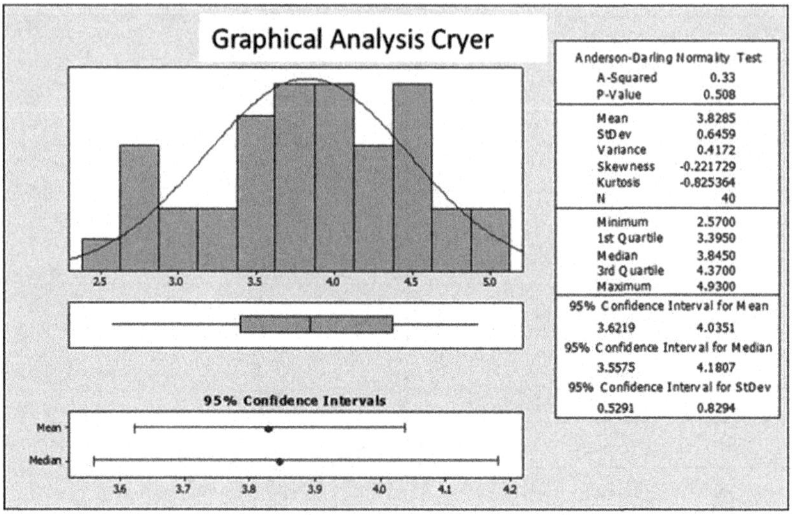

Graphical Analysis for Cryer

FIGURE 5.13
Graphical analysis.

A hypothesis test was performed in order to see if Cryer and Whiner were performing the same taking 30 random samples:

The assumptions are shown in the null and alternate hypothesis:

Ho = (The null hypothesis): The difference is equal to the chosen reference value μ1−μ2=0

Ha = (The alternate hypothesis): The difference is not equal to the chosen reference value μ1-μ2 is not =0

Paired T test for Cryer vs Whiner				
	N	Mean	Std Dev	SE Mean
Cryer	30	82.62	5.2	0.95
Whiner	30	79.7	5	0.91
Difference	30	2.92	0.2	0.04

Paired t-test Cryer vs Whiner

95% CI for mean difference: (1.16, 6.69) T-Test of mean difference = 0 (vs not = 0):

T-Value = 2.90 P-Value = 0.007

FIGURE 5.14
Hypothesis test.

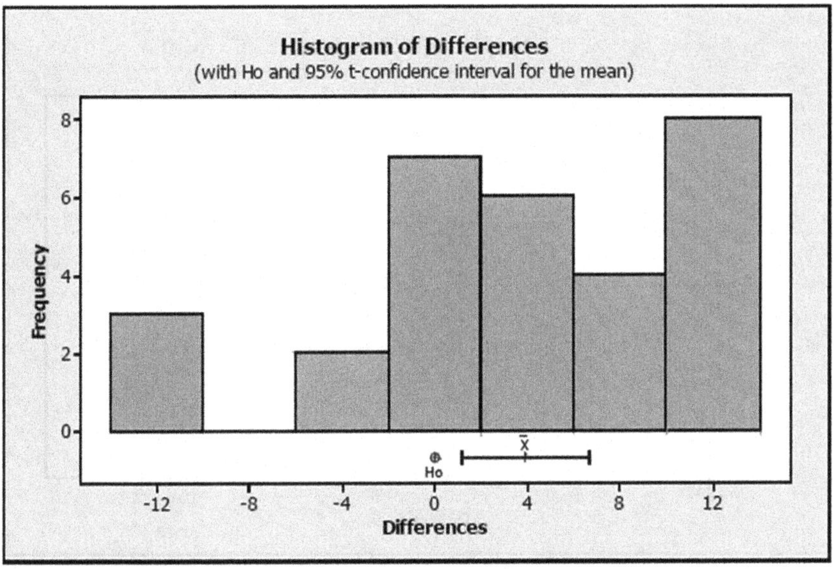

FIGURE 5.15
The histogram of differences *t*-test.

Boxplot of difference *t*-test (Figure 5.16).
Analysis on the hypothesis test:

- Both histograms and box plots show the confidence intervals mean to be greater than zero, which makes it inconsistent with the null hypothesis. Therefore, this should be rejected.

What is the conclusion that can be made about Cryer versus Whiner? What would the suggestion be?

- The conclusion should be to reject the null hypothesis, confirming that there are differences between the two suppliers which affect the amount of rework that is being done by Argo. We can also state that the materials that come from Cryer have outperformed the materials of Whiner.
- Materials that come from Whiner should be stopped being used immediately. And inventory should be checked to see how much of the Whiner material was used so that the factor can get an idea of how much rework will have to be done. This will help prep the Argo to see how to handle the influx of rework that will have to be done and what other expenses are needed such as staff and equipment maintenance.

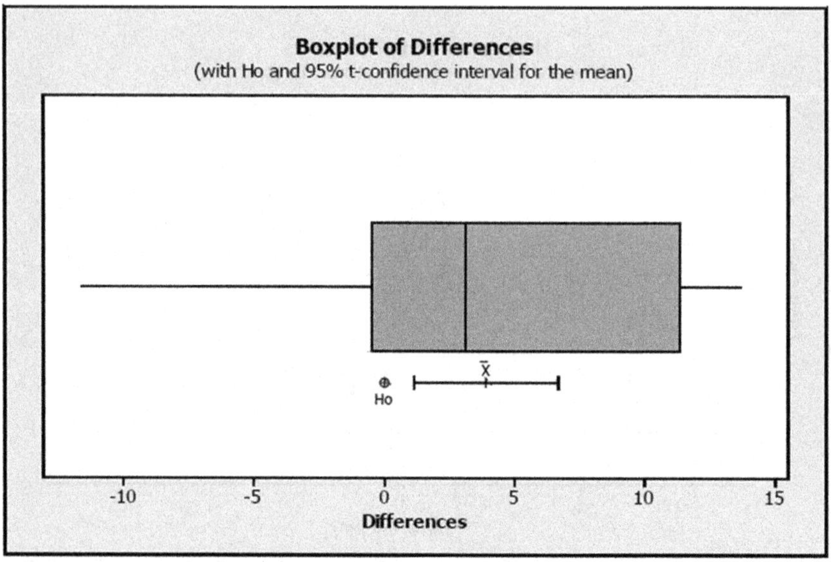

FIGURE 5.16
Boxplot of difference *t*-test.

- Increase the standing order from Cryer for a period of time. Monitor the strict use of Cryer materials to see if the amount of rework will be decreased.
- Cancel any future standing orders with Whiner. Ask the Whiner company for further investigation on their part so that they can correct the issue they have with their materials.

I-MR chart for Cryer:

All of the plots for Cryer's fall with the limits and appear to be stable (Figure 5.17).

Normality chart for Cryer:

The chart looks accurate and the data are normal, indicating that Cryer is performing well (Figure 5.18).

A capability analysis:

A capability analysis was done on Cryer for the 30 random data points on the B side of the line and the process is stable (Figure 5.19).

Gage R and R for Cryer:

- The charts indicate that the gauge is acceptable both in terms of repeatability and reproducibility. Even though there might be a slight variation due to the different operators that are conducting the testing still falls with each other (Figure 5.20).

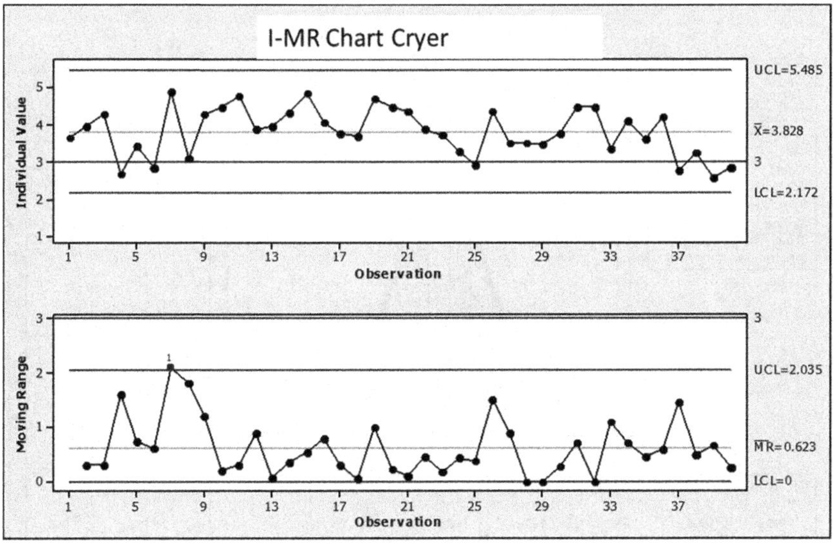

FIGURE 5.17
I-MR chart for Cryer.

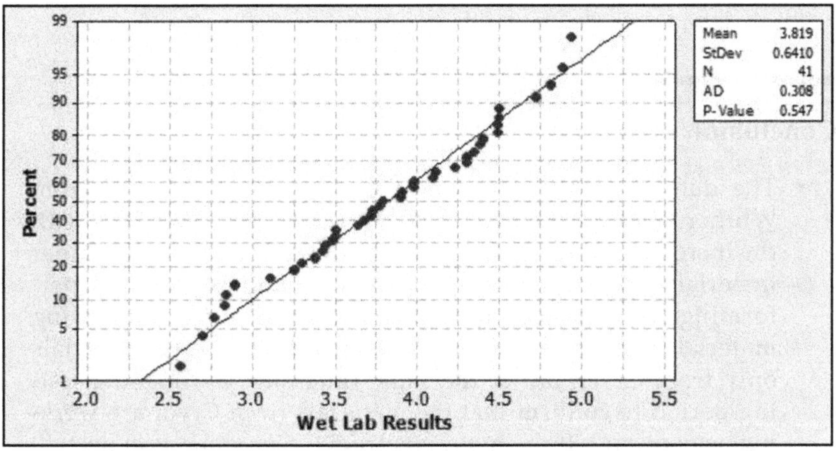

FIGURE 5.18
Normality chart for Cryer.

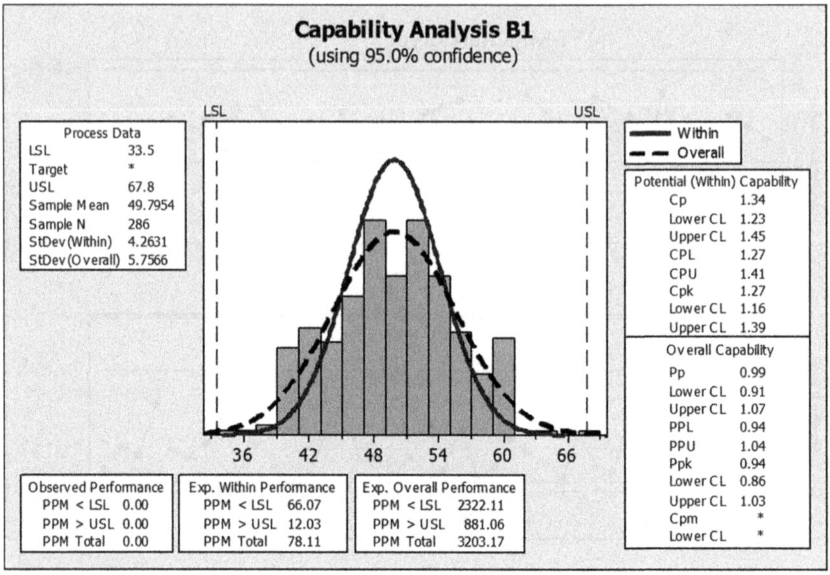

FIGURE 5.19
A capability analysis.

Conclusion of Improve

- The data show that the materials that were obtained from Whiner were the root cause of the issues that Argo had with the increase of rework that had to be done. Even though Whiner materials fell within the criteria, the materials were not performing as expected. It is recommended Argo stop using materials from Whiner immediately and strictly use materials only from Cryer, but at the same time have an ongoing testing period to confirm that the materials from Cryer are working as expected. If the materials from Cryer work as expected, the amount of rework will decrease along with any cost that was associated with the rework process. This will also reduce the number of customers' complaints due to faulty toy product that Argo was putting into the market.

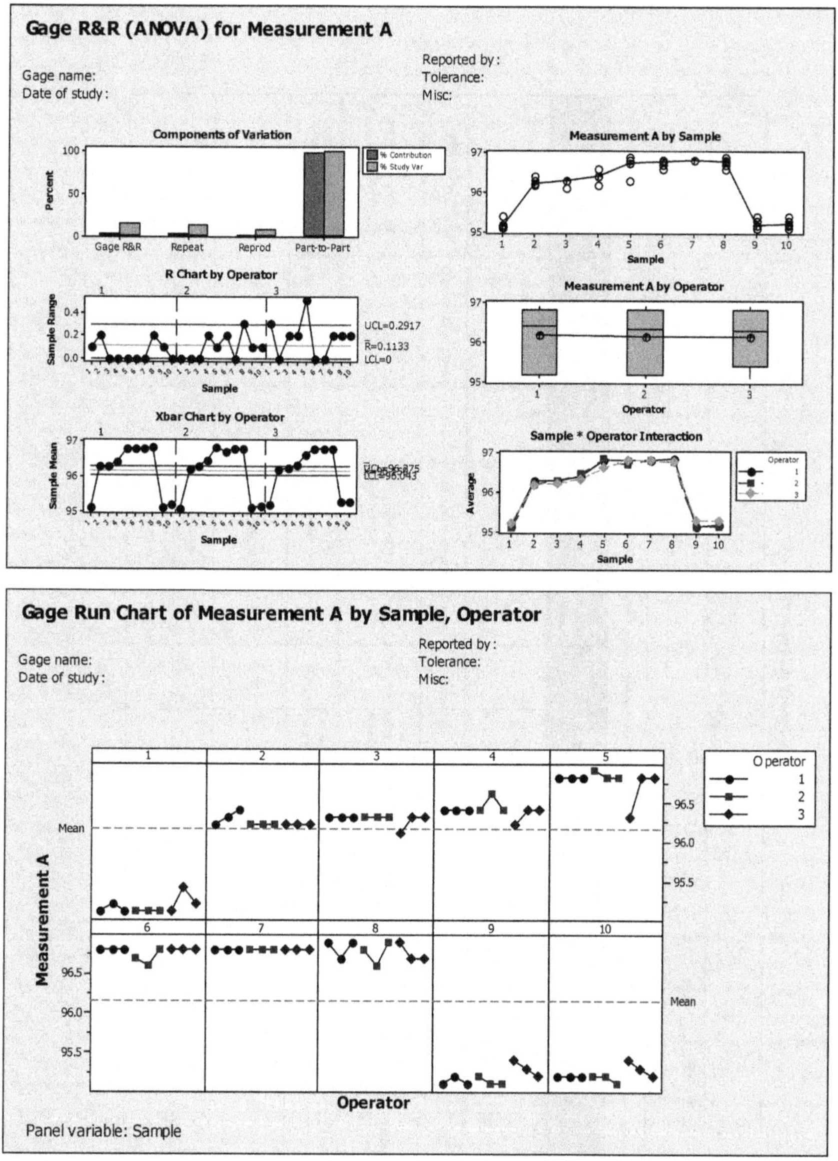

FIGURE 5.20
Gage R and R for Cryer.

Control Plan

Process:	Argo Re-work Reduction	**Preparer:**	Margaret Sanchez	**Page:**	____ of ____
Customer:	Argo Dog Toy Factory	**Email:**		**Reference No:**	
Stakeholder:	Argo	**Phone:**		**Revision Date:**	
Business:	Argo Dog Toy Factory			**Approval:**	

Process	Process Step	CTQ/Metric	CTQ / Metric Equation	Specification/ Requirement (LSL / USL)	Measurement Method	Sample Size	Measure Frequency	Responsible for Metric	Link or Report Name	Corrective Action	Responsible for Action
Check invertory only using Cryer materials	Check stock room for Cryer material and remove all materials from Whiner	Check daily	N/A	Check inventory before and end of shift	Number of toys built		daily	Stock room supervisor		Only count materials from Cryer. Remove all Whinner materials	Supervisor
Order Materials from Cryer	Order Cryer materials	Weekly	N/A	Confirm order from Cryer	Amount of material in stock		weekly	Ordering Department		Order from Cryer only	Ordering Department Supervisor
Receive Materials from Cryer	Check shippment from Cryer	when received (Weekly)	N/A	Inspect shippment frim Cryer	Check amout of materials received verses what was ordered		Upon reciept	Receiving departemnt		N/A	Receiving department Supervisor
Distribute materials to workstations	Take materials to correct area in the factory	Daily		Distraubition of materials	Amount of toys that can be built		per shift	Lead equipment user		N/A	Floor Supervisor
QA check	Check for QA issues	Random	N/A	Check QA on random toys built from different workers and equiment	QA inspections		per shift	Lead equipment user		N/A	Floor Supervisor
Packaging	Packing	Daily	N/A				Daily	Packing Department		N/A	Packing Department
Shipping	Send to Customers/Stores	Daily	N/A				daily	Shipping Department		N/A	Shipping Department

FIGURE 5.21
Control plan.

Control Phase

- Control plan
- Audit checklist
- Conclusion to the project

Control plan (Figure 5.21).
Audit checklist (Figure 5.22).

Conclusion of Project

- Upon looking at the different area of the Argo work process, we have been able to determine what was causing the increase of rework that was being done. We looked at work area, staff, equipment, and raw materials. By looking at the different variables, we have been able to determine that the raw materials from one of the companies used were the cause of the issues with the dog toy

Audit Checklist

Target Area: Argo	Statement of Audit Objective: Reduce Re-work	Auditor:	Audit Date:
Use of Cryer Materials Only		**MS**	5/7/2021
Audit Technique	Auditable Item, Observation, Procedure etc.	Individual Auditor Rating	
Observation	Have all associates been trained to only order from Cryer?	YES	NO
Observation	Is training documentation available for ordering?	YES	NO
Observation	Is training documentation current?	YES	NO
Observation	Are associates wearing proper safety gear?	YES	NO
Observation	Are SOP's available to include QA inspections?	YES	NO
Observation	Are SOP's current?	YES	NO
Observation	Is quality being measured for the Cryer materials?	YES	NO
Observation	Is sampling being conducted in random fashion?	YES	NO
Observation	Are control charts in control	YES	NO
Observation	Are control charts current?	YES	NO
Observation	Is the process capability index >1.0?	YES	NO
Number of Out of Compliance Observations			2
Total Observations			35
Audit Yield			94%
Corrective Actions Required			Yes
Auditor Comments			
Corrective Actions Required			

FIGURE 5.22
Audit checklist.

and creating rework. The increase of the rework also increased the amount of money that had to be used to produce a quality dog toy. The increase of rework also increased materials used, staffing, additional use of equipment, and other resources, which in turn decreased the amount of profits Argo had.

- By taking a closer look at the raw materials that were used, we were able to determine that even though both companies compiled with the regulations put forth by the company, they didn't perform at the same level. Determining that Cryer had a high thread count, it has been found that the materials used for dog toys were more stable than that of Whiner materials. And that all Whiner materials should not be used, effective immediately. This project proved that the use of Cryer materials would be more profitable with less customer complaints. They should give all Cryer materials a testing period to ensure that the materials maintain the same level of thread count to avoid any further rework and complaints from customers.

6

Business Process Management Case Study

Change has a considerable psychological impact on the human mind. To the fearful it is threatening because it means that things may get worse. To the hopeful it is encouraging because things may get better. To the confident it is inspiring because the challenge exists to make things better.

—King Whitney Jr.

The following business process management case study reflects a real-life manufacturing problem with continuous improvement and Lean Six Sigma tools to show how some of the tools are put into place in the real world.

Executive Summary

Whitney Water Co. is a growing product for a large water company. Over the past year, there has been a huge shift in culture due to management changes and employee turnover. There is a new factory manager and he has noticed that the productivity of the facility continues to decrease and the factory is not able to keep up with the demand. Although sales growth has been great, the profit margins are eroding, and the on-time performance is slipping. As the business grows, it gets harder and harder to effectively run the operations. The very important customers have become vocal with their dissatisfaction. As products and services proliferate, it is becoming increasingly difficult for the call center operations to respond in a timely manner. As the operators take longer to provide satisfactory answers to the calls they are on, the incoming calls tend to queue up, and wait times increase. There is some evidence of callers dropping off (Figure 6.1).

Goal Statement: Increase productivity of the factory from 65% to 95% by the third quarter while working on employee satisfaction and morale. Ensure customer demand is at least at 99%.

Methodology: Project Management, Change Management, Data-Based Decision-Making, Leadership, Business Architecture, Lean and Six Sigma, and Innovation and Sustainability.

DOI: 10.1201/9780429184123-6

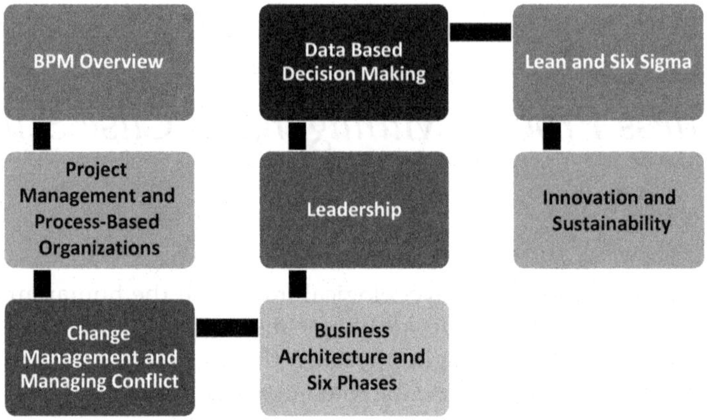

FIGURE 6.1
Business process management.

Key Products:

Water bottles

Call Center Operation:

Ten-hour Monday to Friday call center for product and process inquiries and details including customer complaints

Ten-hour Monday to Friday call center for new orders

Main Customers:

Majority age group between 30 and 50 years old

Employees:

45 professional, technical, and salaried staff and 25 hourly employees

One main facility in Tampa, Florida

Two main suppliers: PiCola and Cpep

Financial Position:

The company is profitable, but the customers are complaining about the on-time rates of the water bottles.

Sales have grown 15% per year over the last five years.

The company has had 25% turnover in employees and 50% turnover in management positions.

Project Management and Process-Based Organizations

A stakeholder identification diagram and analysis was created based on the high turnover of key management positions and employee turnover.

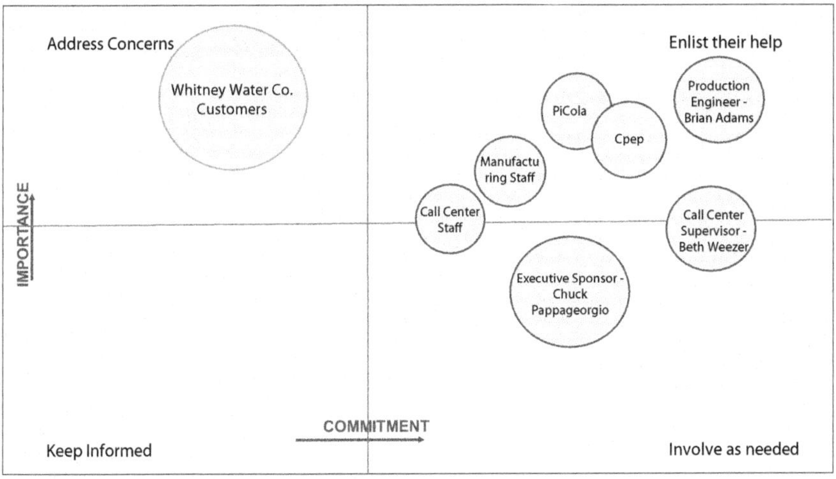

FIGURE 6.2
Stakeholder identification diagram.

The sphere size reflects the sphere of influence by size (Figures 6.2 and 6.3; Tables 6.1 and 6.2).

Conclusion of Project Management and Process-Based Organizations: A stakeholder is any person or organization of persons who can affect or be affected by the actions of the company. It is important to identify their purpose, commitment, importance, and influence using a SWOT analysis.

Strengths and Opportunities (SO) – How can you use your strengths to take advantage of these opportunities?

Strengths and Threats (ST) – How can you take advantage of your strengths to avoid real and potential threats?

Weaknesses and Opportunities (WO) – How can you use your opportunities to overcome the weaknesses you are experiencing?

Weaknesses and Threats (WT) – How can you minimize your weaknesses and avoid threats

Change Management and Managing Conflict

The Change Profile Chart was completed based on the stakeholder identification analysis and SWOT analysis.

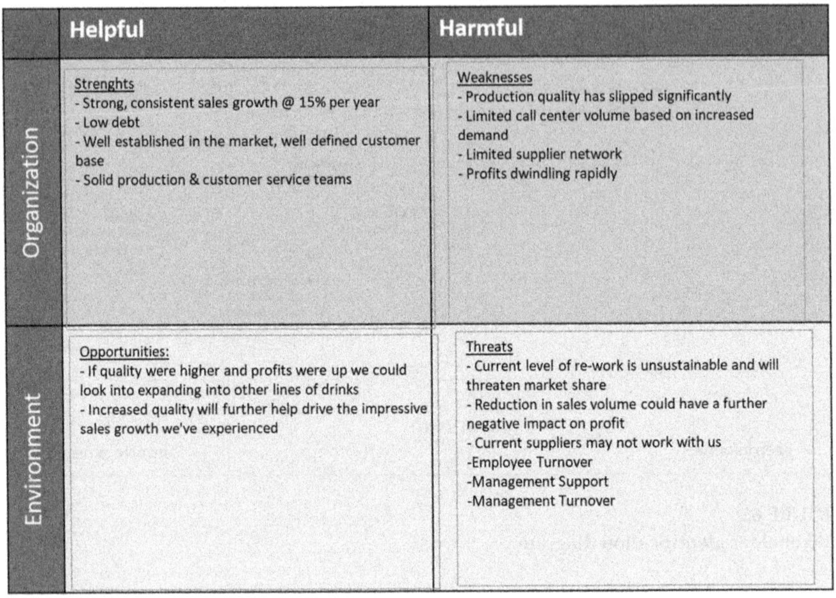
Helpful
Harmful
Organization
Environment
Strenghts
- Strong, consistent sales growth @ 15% per year
- Low debt
- Well established in the market, well defined customer base
- Solid production & customer service teams
Weaknesses
- Production quality has slipped significantly
- Limited call center volume based on increased demand
- Limited supplier network
- Profits dwindling rapidly
Opportunities:
- If quality were higher and profits were up we could look into expanding into other lines of drinks
- Increased quality will further help drive the impressive sales growth we've experienced
Threats
- Current level of re-work is unsustainable and will threaten market share
- Reduction in sales volume could have a further negative impact on profit
- Current suppliers may not work with us
-Employee Turnover
-Management Support
-Management Turnover

FIGURE 6.3
SWOT analysis.

TABLE 6.1
Guiding Stakeholder Questions

Guiding Questions: Stakeholder Analysis
Stakeholder Group
Who are impacted by the change?
Who have an interest in the outcome of the change?
Who are the "owners" of the changes to the business?
Whose role will change from passive to active participation as the changes unfold?
Think of internal and external customer, suppliers, other departments, other project groups, colleagues.
Fill in the name of the group of stakeholders.
Key Individuals
Who is the leader of the group?
Who is the key influencer in the group?
What is their role in the group?
What is their contact information?
Fill in the name of the leader of key influencer.
Engagement Actions
What are the concerns and issues of the stakeholder group with reference to the change?
How do you get the stakeholder group on board? How are you going to involve the stakeholder?
How and when are you going to communicate with the stakeholder?

TABLE 6.2

Stakeholder Analysis

Stakeholder Group	Stakeholder Name	Their Role in Change Effort	Stakeholders' Interests/ Priorities and View on Project	Influence on Program Success	Desired Level of Commitment	Current Level of Commitment	Gap in Commitment	Overview of EngaOverview of Engagement Actionsgement Actions	Program Team Member Assigned	Due Date
Definition:										
Whitney Water Co. Customers		Ultimately this is why we are initiating the change effort to satisfy this stakeholders needs/ demands.								
Production Team	Brian Adams	Collaborate, coach, and ultimately train and implement new processes/ procedures.	Low interest continue to blame quality issues on suppliers.	High	Committed	Aware	High	Document current process, get buy-in from key stakeholders within group. Encorpora te production team into the change management process.	Buddy Lee	Ongoing

(Continued)

TABLE 6.2 (CONTINUED)

Stakeholder Analysis

Stakeholder Group	Stakeholder Name	Their Role in Change Effort	Stakeholders' Interests/ Priorities and View on Project	Influence on Program Success	Desired Level of Commitment	Current Level of Commitment	Gap in Commitment	Overview of EngaOverview of Engagement Actionsgement Actions	Program Team Member Assigned	Due Date
Customer Service Team	Beth Weezer	They'll need to understand the changes we've made and be able to be the "voice" of Argo in dealing with upset customers.	Very interested in helping, just want to see call center volume decrease.	Low	Committed	Aware	Low	Get engaged and keep informed of the changes being made.	Tammy Hays	Ongoing
Suppliers	PiCola and Cpep	Better product quality @ same or less cost than currently spending.	Feel they are providing a quality product at a fair price. Slightly interested.	High	Committed	Aware	Medium	Identify specific quality concerns, work to find win-win solutions for both parties.	Nelly Curtis	Ongoing
Executive Team	Chuck Pappageorgio	We'll need Leo's support and buy-in to obtain resources needed to implement the change effort.	Very interested, quality has become a major concern. High on executive teams radar.	High	Committed	Buy-in	Low	Keep informed of the projects progress. Manage expectations.	Tina Agustiady	Ongoing

Conclusion of Change Management and Managing Conflict: We need to understand the customer and what exactly it is that they want. Then we need to have a continuous flow of production where we aren't interrupting our process with waste. We must pull materials showing when the start and finish of a product is. Each process pulls for the next.

We must always eliminate waste. Waste is any non-value-added activity.

Change involves yourself changing along with others. Change can always be for the better, but no change at all can never lead to better situations. Change management relies on the understanding of why things are done and why people are comfortable. Managing others through change processes is needed to change the status quo. This type of change needs guidance, encouragement, empowerment, and support.

Data-Based Decision-Making

The team wants to know if there is an obvious reason for call times being higher than they need to be. One of the team members suggested using a stem-and-leaf diagram to get a visualization of the variation in call times. Average call times were recorded for the past 70 days. By looking at the numbers below, we can tell there is assignable cause variation. A stem-and-leaf diagram was constructed from the data. The process normality was confirmed. The following was conducted:

1. Construct a stem-and-leaf diagram
2. Answer the question: "Is the process approximately normally distributed?"
3. What is the stem-and-leaf diagram telling you?
4. What would you recommend? (Figure 6.4)

Conclusion of Data-Based Decision-Making: We can tell with data now what is happening with our call volume and if it is affecting the business. We have determined our recommendations.

Leadership

We understand paradigm shifts and different personalities. In summary, there are four main colors:

Project: Whitney Water Co.
Deliverable: Stem & Leaf
Student last name: Pappageorgio

9	5																						
8	0	2																					
7	1	5																					
6	3	4	5																				
5	0	0	0	1	1	2	8																
4	4	4	5	6	6	7	7	7	7	7	7	7	8	8	8	8	8	8	8	9	9	9	
3	2	5	7	8																			
2	0	0	0	0	1	1	2	6															
1	0	1	3	4	5	5	6	6	6	6	6	7	7	7	7	7	7	8	8	9			
0	6																						

Is it reasonably normally distributed? No

If it is not normal, what shape is it? (Use the correct name for this shape, not a description.) Bi-modal

Based on this analysis, what is the next thing you would do? Recommend spending more time with the call center team to understand why we have so many calls that can last upwards of 45 minutes or more.

FIGURE 6.4
Stem-and-leaf diagram.

- Red (adventurer)
- Green (planner)
- Brown (builder)
- Blue (relater)
- Understanding someone's primary behavioral style helps you understand what makes them "tick." A paradigm shift matrix was completed to understand the different behaviors and dynamics in the team so that the team would be successful in the future (Figure 6.5).

Conclusion of Leadership: We can tell with data now what is happening with our call volume and if it is affecting the business. We have determined our recommendations.

Business Architecture and the Six Phases

A solution selection matrix was created based on the ideas to implement based on the six phases (Figure 6.6).

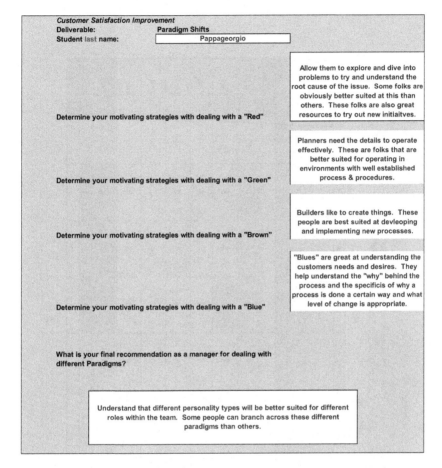

Customer Satisfaction Improvement

Deliverable: **Paradigm Shifts**

Student last name: Pappageorgio

Determine your motivating strategies with dealing with a "Red"

Allow them to explore and dive into problems to try and understand the root cause of the issue. Some folks are obviously better suited at this than others. These folks are also great resources to try out new initiaitves.

Determine your motivating strategies with dealing with a "Green"

Planners need the details to operate effectively. These are folks that are better suited for operating in environments with well established process & procedures.

Determine your motivating strategies with dealing with a "Brown"

Builders like to create things. These people are best suited at devleoping and implementing new processes.

Determine your motivating strategies with dealing with a "Blue"

"Blues" are great at understanding the customers needs and desires. They help understand the "why" behind the process and the specificis of why a process is done a certain way and what level of change is appropriate.

What is your final recommendation as a manager for dealing with different Paradigms?

Understand that different personality types will be better suited for different roles within the team. Some people can branch across these different paradigms than others.

FIGURE 6.5
Paradigm shifts.

Conclusion of Business Architecture and the Six Phases: We can make recommendations and with a solution selection matrix, we are able to know which ideas we truly want to implement.

Lean and Six Sigma

A SIPOC was completed to understand the high-level process and the key inputs, outputs, suppliers, and customers (Figure 6.7).

The next action was to collect data on the quality costs associated with the current system.

Accounting provides the following results for the past year (Figures 6.8 and 6.9).

Solution Selection Matrix							
Project Goal	Please rank each solution for each criteria by usingthe 1-5 Scale as indicated below						
	Very Low (less good)		Moderate		Very High (best)		
	1	2	3	4	5		
Potential Solution (Provide Brief Description)	Potential to Meet Goal	Positive Customer Impact	Cost to Implement (1 = $$$ & 5 = $)	Stakeholder Buy-in	Time to Implement (1 = Long 5 = Quick)	Total Score	Implement? Yes/No
Weighted Criteria	10	9	8	7	5		
Name of Project							
Determine what components of the product are failing	5	4	5	4	5	*179*	Yes
Collaborate with suppliers to find higher quality materials	3	3	2	3	3	*109*	Yes
Implement a intake line to prioritze customer calls by the type of call received	4	4	3	3	4	*141*	Yes
Implement a call back feature on the customer service line	4	4	3	4	5	*153*	Yes
Analyze production practices, determine areas of improvement.	4	4	3	4	3	*143*	Yes
Implement new processes based on research above.	4	4	3	3	3	*136*	Yes
Develop a marketing plan to highlight changes in quality.	3	3	2	2	2	*97*	No
Re-brand product line.	3	3	1	2	1	*84*	No

FIGURE 6.6
Solution selection matrix.

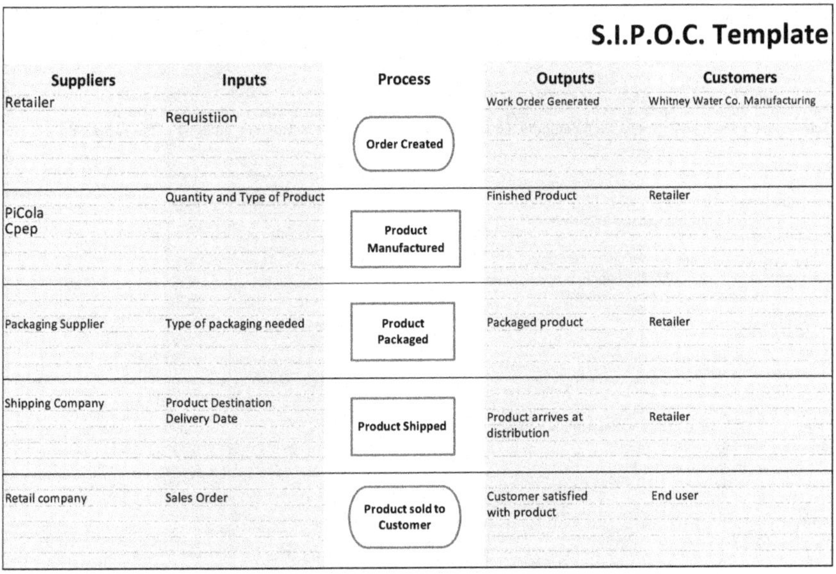
S.I.P.O.C. Template
Suppliers
Inputs
Process
Outputs
Customers
Retailer
Requistiion
Order Created
Work Order Generated
Whitney Water Co. Manufacturing
PiCola
Cpep
Quantity and Type of Product
Product Manufactured
Finished Product
Retailer
Packaging Supplier
Type of packaging needed
Product Packaged
Packaged product
Retailer
Shipping Company
Product Destination
Delivery Date
Product Shipped
Product arrives at distribution
Retailer
Retail company
Sales Order
Product sold to Customer
Customer satisfied with product
End user

FIGURE 6.7
SIPOC diagram.

Raw Materials		**$53,200.00**
Scrap		**$2,416.00**
Incoming Inspection		**$3,700.00**
Improvement projects targeted		**$420.00**
Obsolete products		**$864.00**
Employee certifications for ISO certifications		**$2,097.00**
Corporate audits to verify store compliance with policies		**$3,789.00**
Procedures for employee quality control		**$1,574.00**
Electricity		**$1,196.00**
Inspector fees		**$2,000.00**
Leases for equipment		**$24,020.00**
Scrapped packaging		**$485.00**
Cost for bulk orders		**$297.00**
Free products given to customers with complaints.		**$1,457.00**
Supervisor sign-off/approval for credit card sales		**$348.00**
Business Bureau membership		**$500.00**
Training for incoming inspection		**$368.00**
Loss of business due to customer dissatisfaction		**$6,603.00**
	TOTAL	**$105,334.00**

FIGURE 6.8
Cost analysis.

Customer Satisfaction Improvement
Deliverable: **Cost of Quality**
Student last name: Beard

The categories are: Cost of doing business, Internal Failures, External Failures, Appraisal, and Prevention.
For each of the following, please choose the appropriate COQ category. **Type in your responses. If you are 100% correct when you choose each category, your analysis will match your instructor's analysis of the same data.**

Raw Materials		Cost of Doing Business
Scrap		Internal Failures
Incoming Inspection		Appraisal and Prevention
Improvement projects targeted		Appraisal and Prevention
Obsolete products		Internal Failures
Employee certifications for ISO certifications		Appraisal and Prevention
Corporate audits to verify store compliance with policies		Appraisal and Prevention
Procedures for employee quality control		Appraisal and Prevention
Electricity		Cost of Doing Business
Inspector fees		Appraisal and Prevention
Leases for equipment		Cost of Doing Business
Scrapped packaging		Internal Failures
Cost for bulk orders		Cost of Doing Business
Free products given to customers with complaints.		External Failures
Supervisor sign-off/approval for credit card sales		Appraisal and Prevention
Business Bureau membership		Cost of Doing Business
Training for incoming inspection		Appraisal and Prevention
Loss of business due to customer dissatisfaction		External Failures
A pareto chart of quality costs (do not include chart) shows the business should address which quality cost category first?		Appraisal and Prevention

FIGURE 6.9
Cost of poor quality.

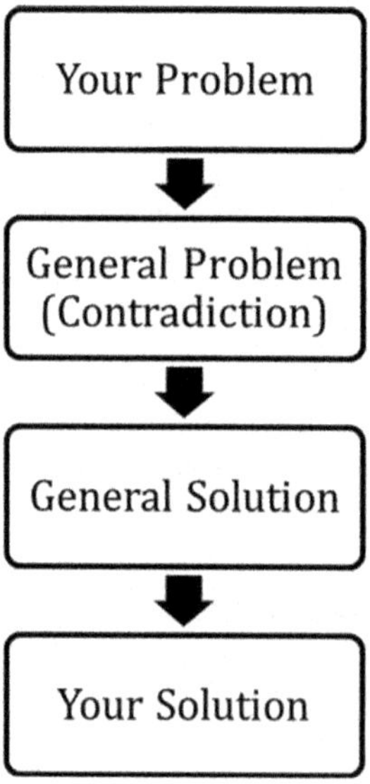

FIGURE 6.10
TRIZ methodology.

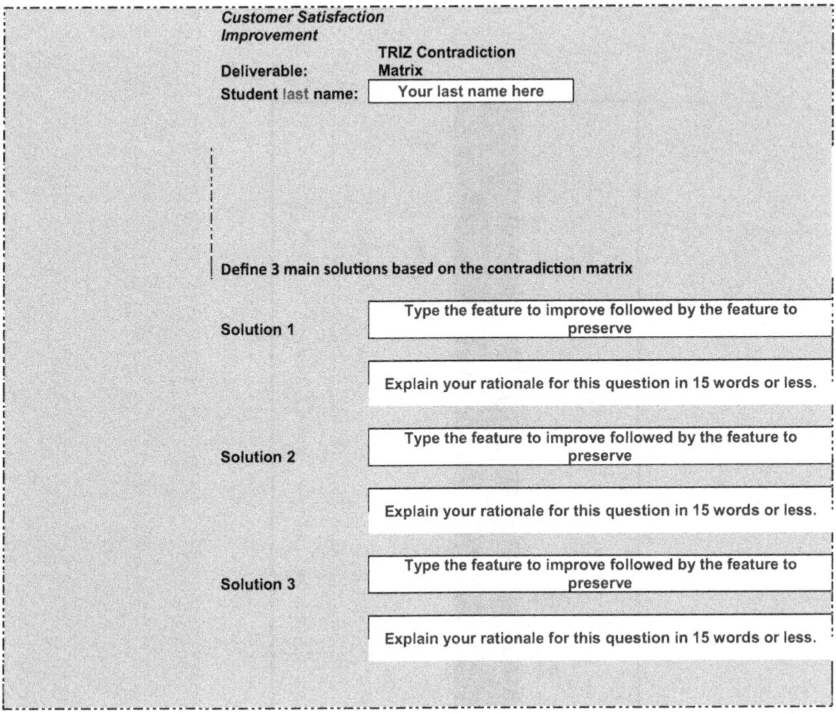

Customer Satisfaction Improvement

Deliverable: TRIZ Contradiction Matrix

Student last name: Your last name here

Define 3 main solutions based on the contradiction matrix

Solution 1
Type the feature to improve followed by the feature to preserve
Explain your rationale for this question in 15 words or less.

Solution 2
Type the feature to improve followed by the feature to preserve
Explain your rationale for this question in 15 words or less.

Solution 3
Type the feature to improve followed by the feature to preserve
Explain your rationale for this question in 15 words or less.

FIGURE 6.11
TRIZ solutions.

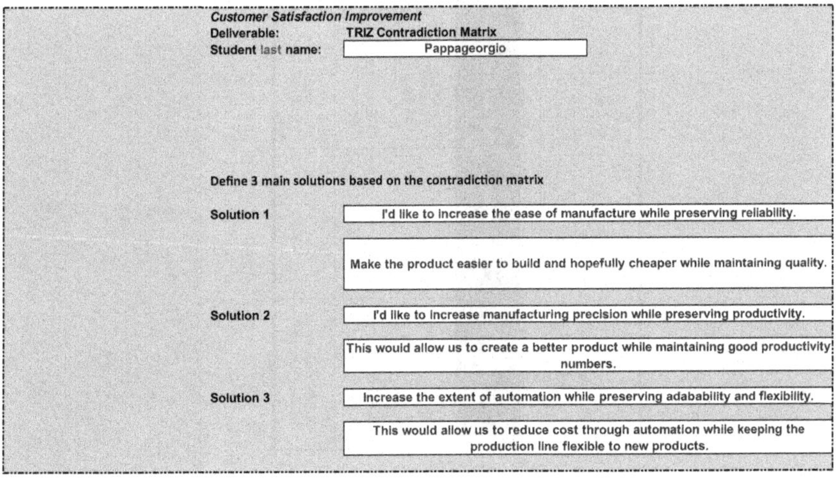

Customer Satisfaction Improvement

Deliverable: TRIZ Contradiction Matrix

Student last name: Pappageorgio

Define 3 main solutions based on the contradiction matrix

Solution 1
I'd like to increase the ease of manufacture while preserving reliability.
Make the product easier to build and hopefully cheaper while maintaining quality.

Solution 2
I'd like to increase manufacturing precision while preserving productivity.
This would allow us to create a better product while maintaining good productivity numbers.

Solution 3
Increase the extent of automation while preserving adabability and flexibility.
This would allow us to reduce cost through automation while keeping the production line flexible to new products.

FIGURE 6.12
TRIZ solution answers.

Process:	Whitney Water Co. Product Lifcycle	**Preparer:**	Brad Beard
Customer:	ARGO	**Email:**	
Stakeholder:		**Phone:**	
Business:		**Owner:**	

Process	Process Step	CTQ/Metric	CTQ / Metric Equation	Specification/ Requirement LSL USL	Measurement Method	Sample Size	Measure Frequency	Responsible for Metric	Link or Report Name
Demand Forecast Generated	1	Accuracy of Demand Forecast	Demand Forecast within 5% of sales	4% 6%	Quarterly sales revenue vs Demand Forecast	All business units	Quarterly	Operations	
Quality products Produced	2	Product Meets Specifications	Product meets specificaitons 95% of the time	94% 96%	Final QA/QC	1/100 units	Daily	Plant Manager	
Quality Customer Service Provided	3	Call Center Volume	Call center volume reduced by 50%	40% 60%	# of calls per month	All call centers	Monthly	Call center supervisor	

FIGURE 6.13
Control plan.

Conclusion
Through a focus working with suppliers to improve the quality of raw goods receivied at Whitney Water Co. manufacturing, improving overall manufacturing processes and a renewed focus on service and employee morale, Whitney Water Co. was able to reduce rework for all water bottles by 17% by Q3 of the year. They are well on our way to a reduction of 25% by June and have significantly reduced the number of customer complaints. Through Paradigm Shifts and cultural change, Whitney Water Co. is well position to experience continued growth within the water business and may expand to other drinks. The Customer Satisfaction Survey showed the employees have increased their confidence and are happy working at the business. The results had a 25% increase in internal Customer Satisfaction.

FIGURE 6.14
Conclusion to project.

The cost of quality categories was filled out in order to understand the key customer worries.

Conclusion of Lean and Six Sigma: We were able to understand our specific suppliers, customers, and how the processes linked to inputs and outputs. We received different costs from accounting and were able to tell what types of cost of poor quality they were referred to.

Innovation and Sustainability

The next action was to complete some Out of the Box thinking. Think of what features you would like to improve and what features you would like to preserve. They defined three areas they would like to concentrate on. They completed a Contradiction Matrix according to a philosophy they found named TRIZ. TRIZ (Theoria Resheneyva Isobretatelskehuh Zadach) is a Russian acronym for the "Theory of Inventive Problem Solving." The term is pronounced as "trēz" or "trees." TRIZ is considered a "left brain engineering" method. The biggest benefit of TRIZ is its ability for developing solutions. It is a structured innovation method for rapidly developing technical and nontechnical concepts and solutions. TRIZ also helps to improve systems by being able to lower costs while increasing benefits. The TRIZ method involves four key steps (Figures 6.10–6.13):

Finally, a control plan was completed in order to maintain the improvements made.

The final conclusion to the project is provided in Figure 6.14.

7

Employing Hoshin Kanri to Drive Lean Implementation and Culture Change

Good leaders set vision, missions, and goals. Great leaders inspire every follower at every level to internalize their purpose, and to understand that their purpose goes far beyond the mere details of their job. When everyone is united in purpose, a positive purpose that serves not only the organization but also, hopefully, the world beyond it, you have a winning team.

—Colin Powell

Company Overview

Carjo Manufacturing is a composite company of my past experiences with various companies that is used in this chapter to illustrate how Lean and Hoshin Kanri should be used together. Carjo Manufacturing produces axle components for the automotive industry. The company's main products include shafts, tubes, and gear housing assemblies. The company's sales for 2020 were $18 million but dropped to $16 million in 2021 because of the market, and sales for 2022 are expected to be down again, slightly, from 2021. Therefore, in order to survive, Carjo must implement Lean to drive out wastes and improve its profit margins.

Management understands the change in the manufacturing environment and sees the need to change in order to survive. In addition, Carjo's senior leadership has heard about the success of Lean and has driven some Lean implementation. The problem has been that the company has seen only modest gains from implementing Lean. The senior leadership team expected more dramatic gains from the Kaizen events that had been held. The senior leadership team consists of the following:

- CEO
- Director of Operations

DOI: 10.1201/9780429184123-7

- Director of New Business
- Director of Marketing
- Director of Engineering
- Director of Quality
- Director of Finance

Carjo's operations consist of three plants:

1. The first plant manufactures the shaft for the gear assembly.
2. The second plant machines and assembles the tube.
3. The third plant machines the gear housing and then assembles the gear assembly.

The three facilities are on the East Coast and in close proximity, but in different states. None of the plants are large enough to contain all the necessary operations that they currently operate.

The production processes for the final product, the gear housing assembly, involve Plants 1–3. Transportation time is two days by truck from Plant 1 to Plant 3, One and a half days by truck from Plant 2 to Plant 3, and one day by truck from Plant 3 to the customer.

The raw materials at each plant are supplied by various vendors. Deliveries of castings, bearings, and bushings occur once a week to Plant 1 for the production of the tubes. Plant 2 receives deliveries of shaft castings every two weeks. Plant 3 machines the gear housing and performs the final assembly. For this process, Plant 3 receives shipments twice a week from Plant 1 (tubes), Plant 2 (shafts), and the gear supplier. Plant 3 then makes daily deliveries to the customer.

The main operations, including Carjo headquarters, are housed at Plant 3. This is where the main Production Control Department resides. Carjo receives its customers' 90/60/30 day forecasts and enters them into the materials requirements planning (MRP) system. In turn, Carjo issues a 90/60/30 day forecast to all of its suppliers (of tube castings, bearings, bushings, shaft castings, and gears). Carjo secures its raw materials using a weekly faxed order release to its suppliers. Internally, the Production Control Department generates MRP based, weekly departmental requirements based on customer orders. Based on this information, the Production Control Department issues daily build schedules to Plants 1–3. Daily shipping schedules are also issued to the shipping departments at Plants 1–3.

In terms of the macro level, the process information is shown in Figure 7.1.

The macro-level value stream map is shown in Figure 7.2.

Based on their macro-level value stream map, the senior leadership team at Carjo, which consists of the CEO and Directors, could see that

Plant 1:

Raw material is shipped in by truck
Shipments are received once a week, usually on Fridays to ensure the raw material is received prior to the start of the next production week
Shipments from the casting supplier and bearing supplier take 1 day
Shipments from the bushing supplier take 2 days
Lead time is 8.8 days
No changeover is required
Process reliability is 84%
Yield is 80%
First pass yield is 63%
Observed inventory:

- 7,802 tubes (in receiving)
- 8,986 bushings (in receiving)
- 6,329 bearings (in receiving)
- 275 tubes machined and washed (ready for bearing press)
- 30 final products waiting for packaging
- 3,870 tubes packaged and waiting for shipment to the customer

Tubes are shipped to Plant 3 by truck
Deliveries occur twice per week and take 2 days

Plant 2:

Raw material is shipped in by truck
Shipments are received twice a month and take 2 days
Lead time is 16.2 days
The plant follows the same process for the two shafts produced
Changeover is required on most of the processes for the two different types of shafts
Process reliability is 90%
Yield is 96%
First -pass yield is 87%
Observed inventory:

- 11,650 shaft castings (in receiving)
- Various levels of WIP between the processes (shown in detail in process level maps later in Figure 7-3)
- 3,115 shafts packaged and waiting for shipment to the customer

Finished shafts are shipped to Plant 3 by truck
Deliveries occur twice per week and take 1 ½ days

Plant 3:

All material is shipped in by truck (from Plant 1, Plant 2, and Gear Supplier)
Shipments are received twice a week
Shipments from Plant 1 and the gear supplier take 2 days
Shipments from Plant 2 take 1 ½ days
Lead time is 9.4 days
No changeover is required

FIGURE 7.1
Process flow information for Carjo Manufacturing Co.

Process reliability is 95%
Yield is 88%
First pass yield is 80%
Observed inventory:

- 4,492 shafts (in receiving)
- 2,783 tubes (in receiving)
- 2,314 gears (in receiving)
- Various levels of WIP between the processes (shown in detail in process level maps later in Figure 7-4)
- 903 gear housing assemblies, packaged and waiting for shipment to the customer

Finished assemblies are shipped to the customer by truck
Deliveries occur daily and take 1 day

FIGURE 7.1
Continued

they have significant problems. Coupled with the current market conditions, Carjo must improve its processes to ensure long-term business viability. Lean has provided some benefits but has not had the overall impact on the business that the leadership team envisioned. There is a disconnect between the improvement activities and where the organization must go strategically. Carjo has realized modest gains from implementing Lean techniques. However, Carjo can achieve more significant results by tying their improvement efforts to their strategic long-term objectives.

The leadership team at Carjo has identified a disconnect between the long-term strategy of the business and the improvement efforts that have taken place. The company leaders have decided to use Hoshin Kanri to flow down the long-term vision of the organization and use it as the strategy for business process excellence.

Developing Carjo's Hoshin Strategic Plan Summary

The senior leadership team at Carjo Manufacturing scheduled a leadership retreat to develop the long-term vision of the organization. The team met for three days and developed the following four strategic goals for the organization:

1. Develop world-class new business/marketing
2. Expand product offerings
3. Implement cost reduction projects
4. Implement Kaizen projects

The senior leadership team then used the strategic goals to develop three core objectives for the organization:

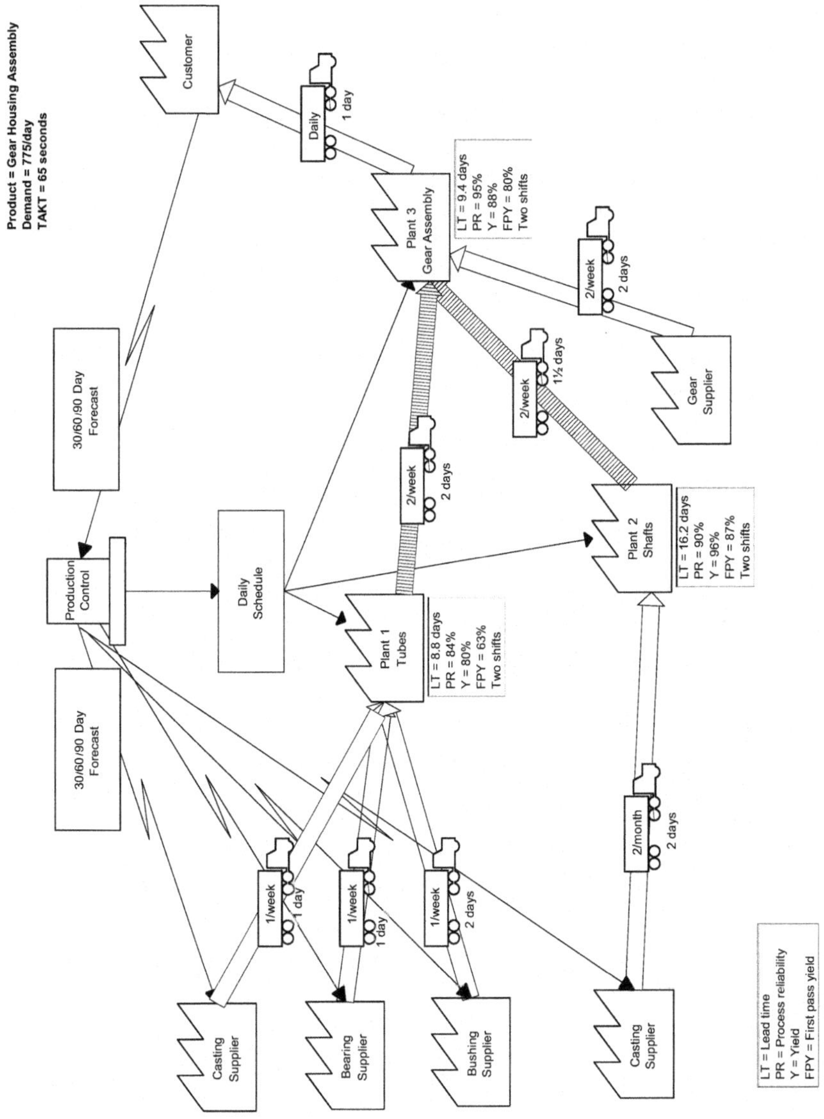

FIGURE 7.2
Carjo's macro current state map.

1. Improve financial returns by 5% by 2027
2. Grow sales by $3M by 2025
3. Achieve world-class supplier status by 2023

The leaders then reviewed the relationships between their strategic goals and their core objectives to ensure all of their strategic goals were being addressed properly. Next, they identified the appropriate metrics that would tie directly to their core objectives. The metrics needed to be quantitative (and not qualitative) to indicate whether the process improvements have an impact on the overall organization and are trending in the right direction. Finally, the senior leaders assigned ownership of the core objectives to specific member(s) of their team, as shown in Figure 7.3 (and refer back to Figure 3.5, which provides a blank template).

Developing Carjo's Hoshin Plan Summary

The leadership team next focuses on developing the Hoshin Plan Summary: see Figure 7.4 (and refer back to Figure 3.6, which provides a blank template). As you can see, the strategic goals and owners have been carried down from the company's Hoshin Strategic Plan Summary (refer back to Figure 7.3). The short-term and long-term goals must again tie back to the measures outlined in the Hoshin Strategic Plan Summary. The

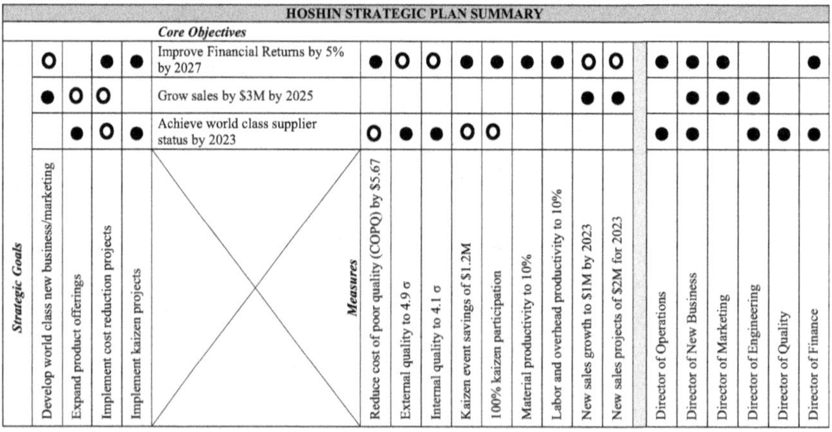

FIGURE 7.3
Carjo's Hoshin Strategic Plan Summary.

HOSHIN PLAN SUMMARY								
Strategic Goals	Mgt. Owner	Goals		Implementation Strategies	Improvement Focus			
		Short Term	Long Term		Safety	Quality	Delivery	Cost
Develop world class new business/ marketing	Director of New Business	$1M by 2023	$3M by 2025	Grow sales				●
Expand product offerings	Director of Engineering	$2M for 2023	$5M for 2025	New sales projects		○		●
Implement cost reduction projects	Director of Operations	$1.2M annual savings	$4M annual savings	Improve financial returns	○	●	●	●
Implement kaizen projects	Director of Operations	100% participation	100% participation	Achieve world class supplier status	●	●	○	●

FIGURE 7.4
Carjo's Hoshin Plan Summary.

leaders at Carjo decided to have their short-term goals focus on improvements in the next year and the long-term goals three years out.

The next big decision for Carjo's leaders was to determine what their implementation strategies would be based on their strategic goals. For example, in order to develop world-class new business/marketing, the leaders determined their implementation strategy would be to grow sales.

Then based on the implementation strategy, the leadership team determined which improvement focus area was impacted. Using the four improvement focus areas, the leadership team at Carjo outlined the impact of each strategic goal: again, see Figure 7.4. Based on this information, the leadership team could clearly see that two strategic goals impacted all four improvement focus areas. These two strategic goals are (1) implement cost reduction projects and (2) implement Kaizen projects, because they show a relationship with each improvement focus.

This practice helped Carjo's senior leaders prioritize their implementation strategies. Because two of the implementation strategies have a significant impact, these should be the company's highest priority.

Now that Carjo has developed its Hoshin Strategic Plan Summary, its Hoshin Plan Summary, and its implementation strategies, each department should utilize current-state value stream mapping to identify improvement opportunities relating to the implementation strategies developed to this point.

Takt Time

Prior to starting the mapping process, the process improvement team first needs to understand its customers' requirements. It is vital for the process improvement team to understand the customer requirements; however,

Calculating Takt Time					
8	Hours =	480	Minutes (Based on standard work shift)		
		-30	Minutes (Break Time)		
		-10	Minutes (Wash Time)		
		-0	Minutes (Clean-up)		
		-5	Minutes (Team Meetings)		
	Total	435	Available Minutes per Shift		
435	Minutes Available x 60 =		26100	Seconds per shift	
26100	Seconds Divided by	775	pcs/shift =	67	Seconds
	Takt Time =	67	Seconds per Piece		

FIGURE 7.5
Carjo's takt time.

every person in the organization should also understand the customer requirements. *Takt time* is the frequency with which the customer wants a product. In other words, it is how frequently a sold unit must be produced. You can calculate this time by dividing your available production time in a shift by your customer demand for products that are made during a given shift. Takt time is usually expressed in seconds. However, depending on the product and the industry, takt time can be expressed in hours, days, or weeks.

For the gear housing assembly at Carjo, the daily customer demand is 775 units. All three plants run a two-shift operation. Therefore, takt time can be calculated as shown in Figure 7.5.

Takt time is 67 seconds per gear housing assembly. This information is critical as we gather our information for the current-state maps. All process operations must be less than takt time.

Current-State Maps for Carjo Plants

The next step for the continuous improvement team is to create current-state value stream maps for all of their facilities. This will help the team understand their baseline processes. Figures 7.6–7.8 show the current-state maps for Plants 1–3, respectively. These provide the baseline for process improvements.

Now that Carjo's leaders have a picture of their company's current state and understand their customer demand, they can begin to identify areas

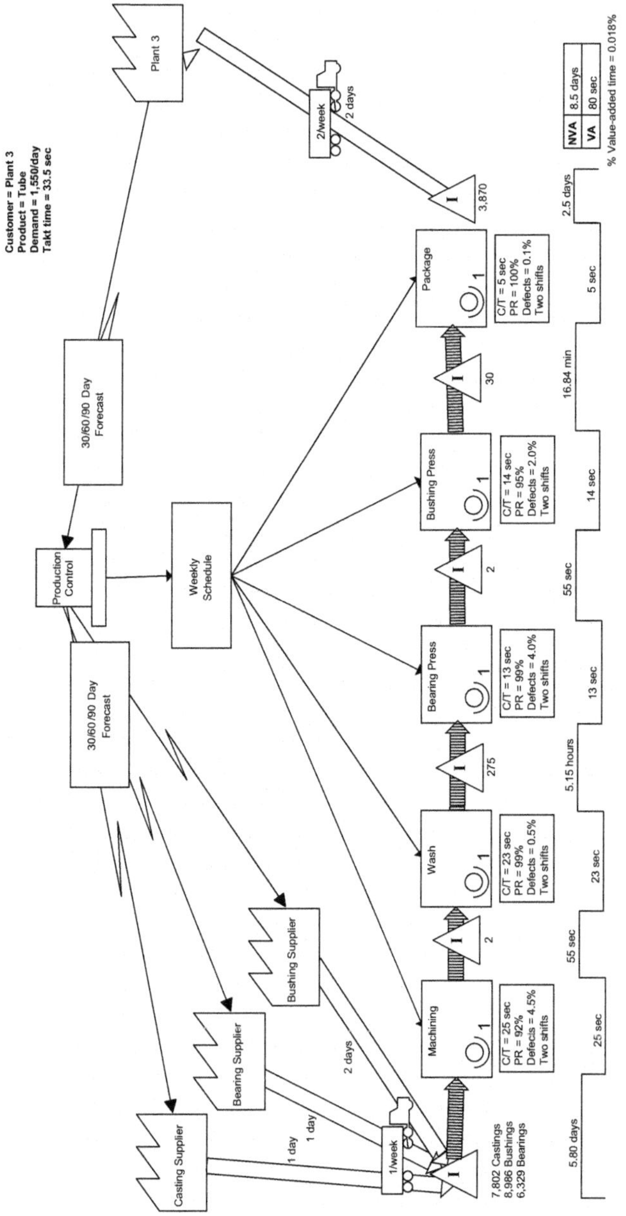

FIGURE 7.6
VSM for the tube process in Plant 1.

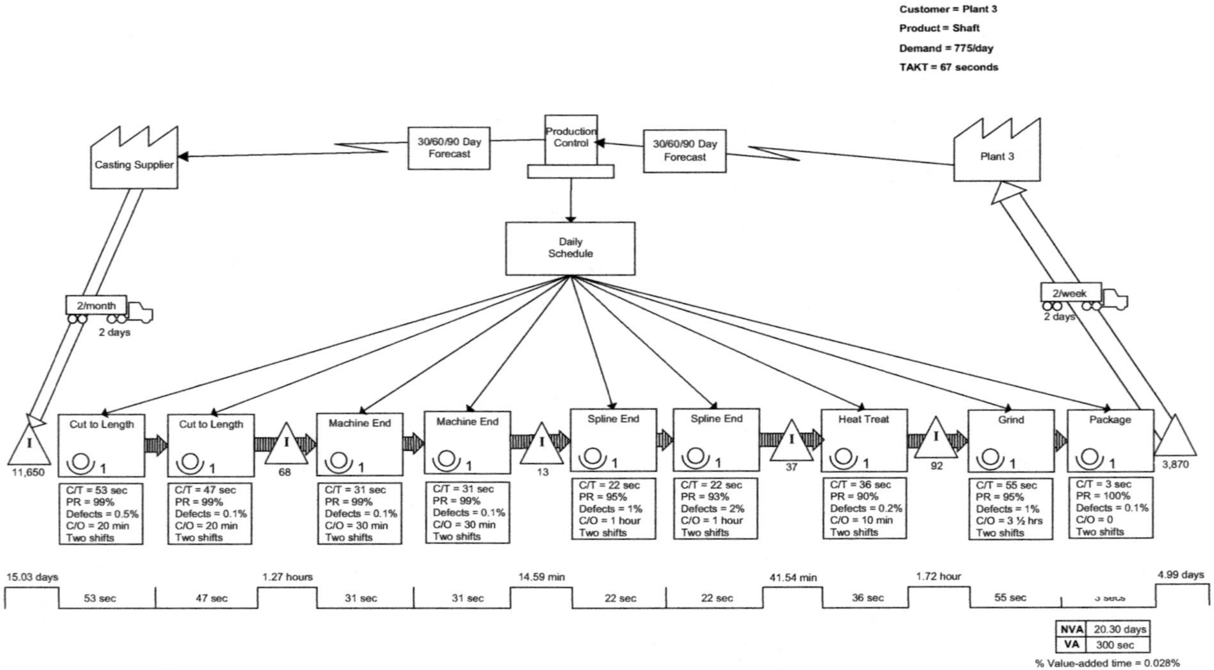

FIGURE 7.7
VSM for the shaft process in Plant 2.

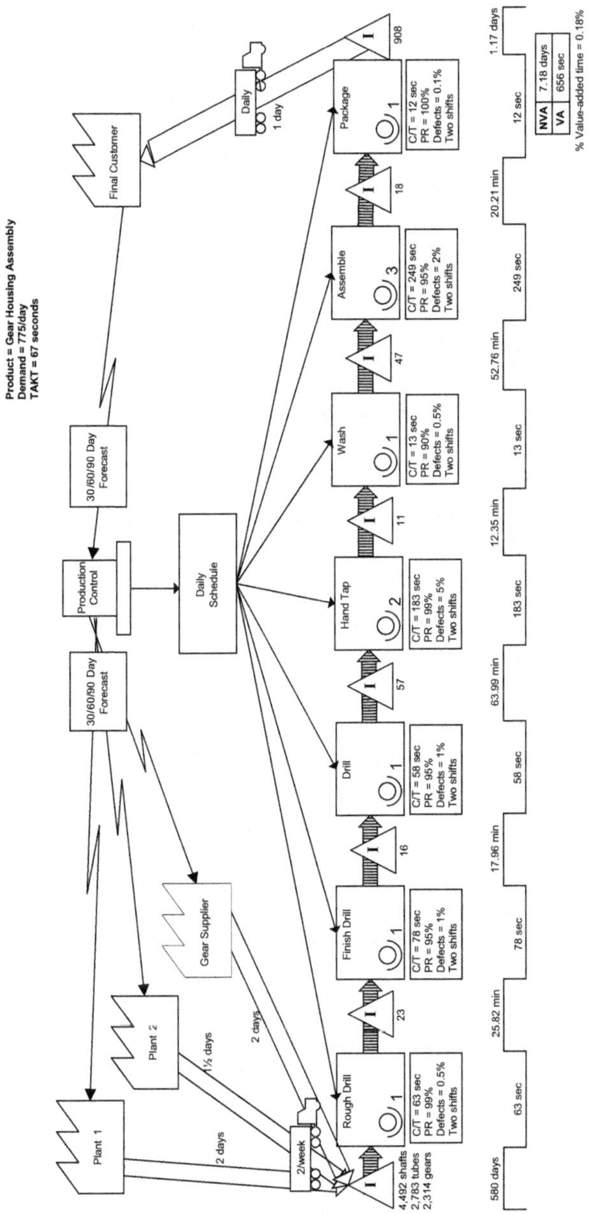

FIGURE 7.8
VSM for the gear housing process in Plant 3.

Waste	Description
Defects	Poor quality of products requires production of additional products (which causes overproduction) to replace the defective parts, and it creates an inventory of unusable products. Poka-yoke techniques can help prevent defects from moving down the production line.
Overproduction	Producing more than what the immediate internal or external customer needs. Overproduction requires additional space, material handling, and storage that otherwise would not be needed.
Waiting	Time spent waiting for materials. This is typically caused by unbalanced production lines.
Non-utilized talent	Non-utilized talent is the eighth waste, which was added later. It represents the need to empower and involve employees. Organizations later recognized the wastes that stemmed from underutilizing their employees talents, skills, and knowledge.
Transportation	Transportation does not add value because it does not contribute to transforming the final product. Point-of-use techniques can help minimize transportation waste.
Inventory	Inventory not immediately needed by the customer. This is typically caused by push scheduling.
Motion	Wasted motion includes double handling, reaching for parts, and stacking parts, to name a few examples. Point-of-use techniques can help eliminate wasted motion.
Excessive processing	Extra or over processing waste can be caused by poor tool or product design.

FIGURE 7.9
Eight forms of waste.

for improvement. The current-state maps and takt time are essential to pinpointing bottlenecks that hinder flow and wastes.

There are several reoccurring issues with the processes in Plants 1–3 that lead to excessive wastes. Waste is any process or operation that adds cost or time and does not add value. The seven original of wastes are described in Figure 7.9. An eighth waste of unused creativity should also be considered: we need to focus on our people and tap into their intellectual creativity. These eight forms of wastes are commonly referred to by their acronym of DOWNTIME (defects, overproduction, waiting, non-utilized talent, transportation, inventory, motion, excessive processing) as reflected in Figure 7.9.

Analyzing Waste in Carjo's Plant #1

Figure 7.10 shows the waste in the tube operations processes in Carjo's Plant 1, using Kaizen bursts (in the shaded areas). As you can see:

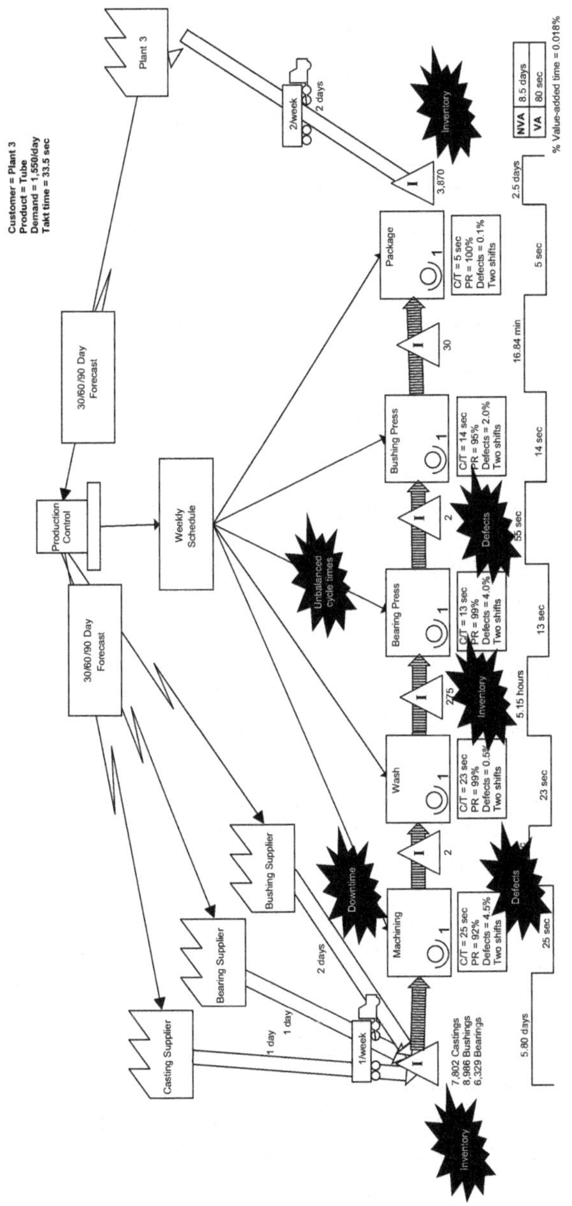

FIGURE 7.10
VSM with wastes for the tube process in Plant 1.

- The cycle times for all of the processes are considerably less than takt time.
- The value-added time is 80 seconds.
- However, the total lead time is 8.5 days, which calculates to a 0.018% value-added time for Plant 1.

The Toyota Production System (TPS) is considered best-in-class for implementing Lean techniques. TPS has several processes that have a percent value add time as high as 35%. This is world-class performance. Companies typically range in the 3%–5% range for percent value-added time when first beginning the Lean journey. In comparison, Plant 1 is operating at an extremely poor level of percent value-added time, which indicates there is considerable opportunity for improvement.

Analyzing Waste in Carjo's Plant #2

Figure 7.11 shows the current state map for the shaft manufacturing process in Carjo's Plant 2; again, the wastes in the processes are shown using Kaizen bursts:

- All of the processes are operating well under takt time.
- The value-added time is 300 seconds, while the total lead time is 20.3 days.
- Therefore, the value-added time for Plant 2 is 0.028%.

The performance in Plant 2 is again well below the industry average.

Analyzing Waste in Carjo's Plant #3

Finally, Figure 7.12 shows the current state map for Carjo's Plant 3, which machines the gear housing and performs the final assembly; again, it shows the waste in the processes using Kaizen bursts. Here's what Figure 7.12 reveals about Plant 3:

- The value-added time is 656 seconds.
- The total lead time, however, is 7.2 days.

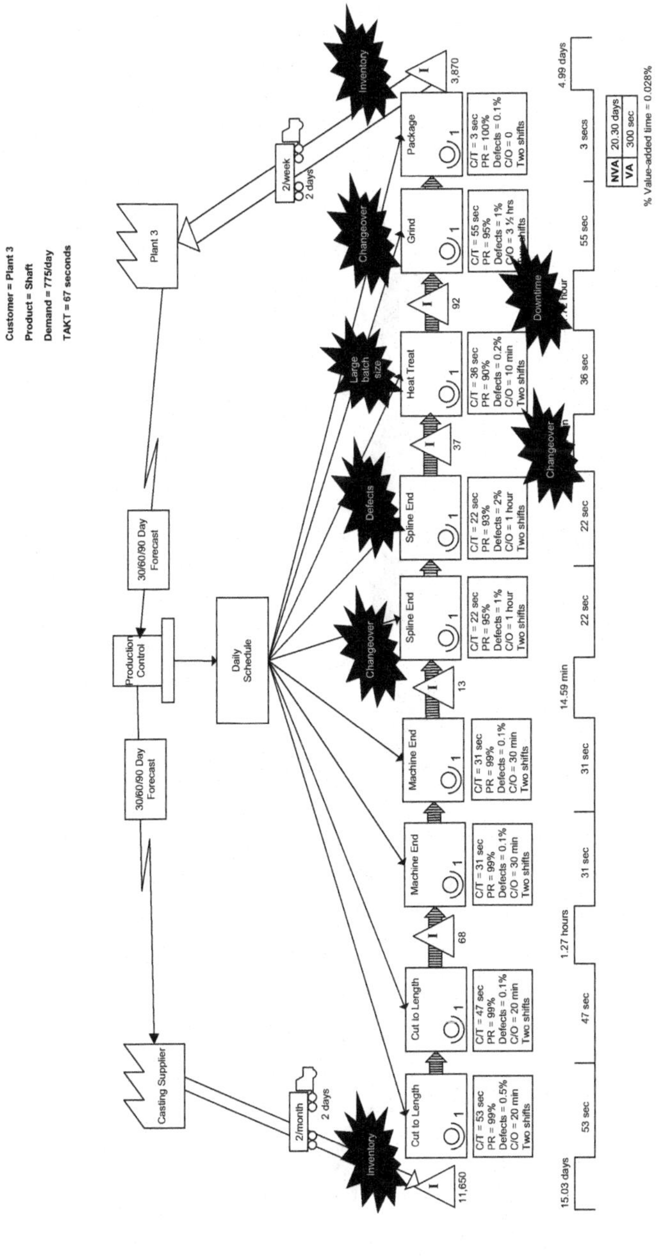

FIGURE 7.11
VSM with wastes for the shaft process in Plant 2.

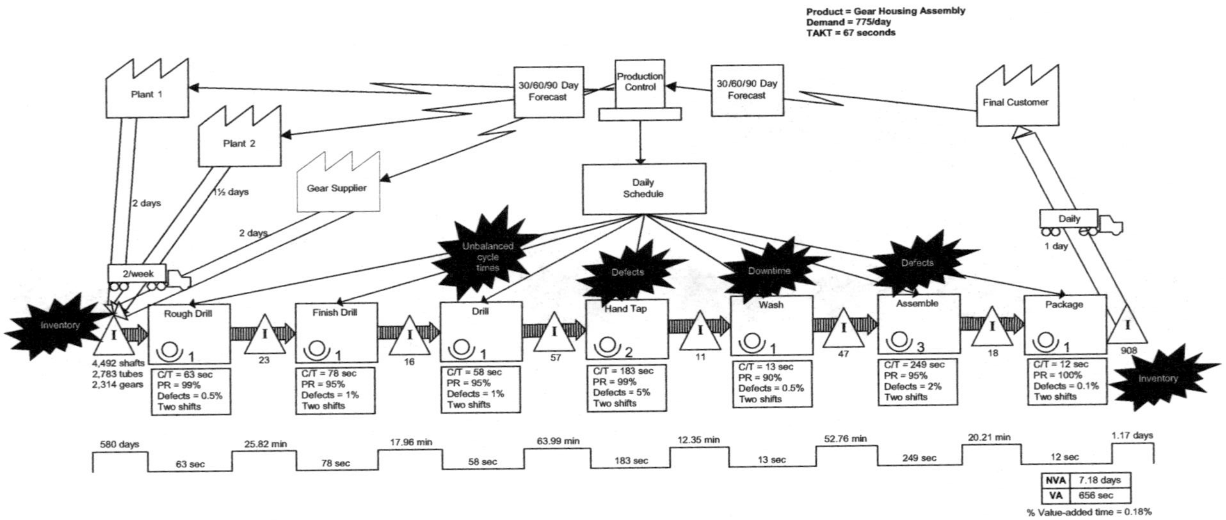

FIGURE 7.12
VSM with wastes for the gear housing process in Plant 3.

- That calculates to 0.18%% value-added time.

Plant 3 has the highest percent value-added time of the three facilities, but it is still an extremely low percent.

In several cases, operators are assigned to a machine that is operating considerably below takt time, which is causing inventory to build up. This leads to the waste of overproduction and waste from idle/waiting time.

Given the takt time of 67 seconds, there are three operations in the gear housing assembly process that cannot be completed within this time. These are bottleneck operations and cause other operations to wait. In addition, the long cycle times are leading to overtime to produce enough product for shipments, which increases production costs.

Waste Elimination

In addition to the problems at Plant #3, there is considerable inventory at each facility. The inventory exists mainly at the beginning and end of the process due to infrequent deliveries. But there is also inventory built up as work-in-process throughout each facility caused by the lack of balance between operations.

For example, in Plant 2, the long changeovers for grind and spline operations are a significant disruption and waste in the process. The grind operation requires a three and a half hour changeover. Each spline machine requires a one-hour changeover.

So where should Carjo begin its Lean implementation? The company's previous method was to have each facility decide on the process improvement activities themselves. The Kaizen event selection was up to the Operations Manager at each facility. No formal method existed to select the projects. The current-state value stream maps were used to drive the improvement project identification. But there was no clear method for prioritization of the projects. Previous projects resulted in gains, but in the overall process at the system level, these gains did not have a big impact.

Now that Carjo's leaders have developed their strategic goals using Hoshin Kanri and understand their current situation using value stream mapping, you can begin the process improvements. At this stage, however, it is critical for you to consider the organization as a system of integrated processes. Some improvement efforts will only impact small portions of the business while other efforts can impact the entire business.

Developing Carjo's Hoshin Action Plan

Recall that the leadership team at Carjo selected the strategic goal of implementing cost reduction projects because of this goal's impact on safety, quality, delivery, and cost. The department managers reviewed their current-state value stream maps (refer back to Figures 7.10–7.12) with their teams to identify improvement opportunities for cost reduction projects. And the manager of the tube line noted a high internal defect rate in the tube line for oversize bushing bores (refer to Figure 7.10).

The oversize bushing bores currently account for 4.3% of the total product costs. This background information is key to understanding and developing the situation summary for Carjo's Hoshin Action Plan. There should be a compelling reason that is linked to the strategic objective based on quantitative data. Without the clear linkage, you may be focusing on improvements that do not impact the overall goals of your organization and, therefore, may be suboptimal improvements.

In addition, the defect had a current internal quality of 56,704 parts per million (PPM). The Department Manager, Leah, selected this project as part of the company's cost reduction strategy. Using their current baseline, the team developed short-term and long-term goals for the project.

The team had been working for several months to reduce the defect rate, but it was still not realizing significant gains. The improvement team could not quantify the key variables that affect the product performance. Therefore, the team decided to pursue this defect issue as a Six Sigma project.

The targets and milestones developed for this project follow the Six Sigma methodology. In this example, the team has outlined its target actions as conducting a measurement systems analysis (MSA) and a process failure modes and effects analysis (PFMEA) with expected completion dates. Figure 7.13 shows the Hoshin Action Plan developed by the team.

The department as a team goes through each core objective to identify implementation strategies that will impact the overall organization. The team develops a Hoshin Action Plan for each of these strategies.

Developing Carjo's Hoshin Implementation Plan

Next, each of these strategies is cascaded down to the Hoshin Implementation Plan, as shown in Figure 7.14. The team at Carjo lists each

Hoshin Action Plan		
Core objective: Implement cost reduction projects **Management owner:** Leah **Department:** Tube line	**Team:** Krista, Brian, Brooke **Date:** 9/14 **Next review:** 10/15	
Situation summary: Internal defects are currently 4.3% of total product cost.		
Objective: Implement cost reduction projects to improve financial returns.		
Short-term goal: Reduce internal quality rate to 32,798 PPM **Long-term goal:** Reduce internal quality rate to 28,154 PPM	**Strategy:** Six Sigma project on oversize bushing bore	**Targets and milestones:** Conduct measurement systems analysis by 9/21 Conduct failure modes and effects analysis by 10/2

FIGURE 7.13
Carjo's Hoshin Action Plan.

Hoshin Implementation Plan

Core objective:
Implement cost reduction projects

Management owner:
Leah

Date:
9/14

Strategy	Performance		Schedule and Milestones											
			Jan	Feb	Mar	Apr	May	June	July	Aug	Sept	Oct	Nov	Dec
Six Sigma project on oversize bushing bore	Target	28,154 PPM	57,304	54,654	52,004	49,354	46,704	44,054	41,404	38,754	36,104	33,454	30,804	28,154
	Actual	31,652 PPM	54,007	54,900	51,538	48,128	42,860	43,957	38,100	37,905				
Kaizen participation	Target	100%	10%	20%	30%	40%	50%	60%	70%	80%	90%	100%	100%	100%
	Actual	70%	12%	23%	31%	42%	55%	58%	74%	82%				

FIGURE 7.14
Carjo's Hoshin Implementation Plan.

implementation strategy in the left-hand column. In order to determine the target improvements, the team revisits the current-state maps for this process. The current state map provides the baseline for the improvement strategy. By understanding the current performance level, the team determines the target improvement.

Once the team members determine their target improvement, they break down this improvement into monthly improvement targets. This enables the team to manage the project and monitor trends. Following the example of the Six Sigma project to reduce oversize bushing bores, the team breaks down the improvement in internal PPM by month. The team can then monitor if they meet their targeted PPM reduction each month.

The team should present this chart to its senior leadership team on a monthly basis. It provides a high-level team picture of current improvement activities. The team can also highlight the projects that are on track by coloring the background green for "project on track" and red for "projects that are not meeting the specified target." In a quick glance, anyone can then see the status. However, since this book is in black and white this color scheme does not work as well. Therefore, in Figure 7.14, the projects that are not meeting the set target are shaded in gray and projects that are on track have no background fill.

Preparing Carjo's Hoshin Implementation Review

The final form is the Hoshin Implementation Review, which is shown in Figure 7.15. This is where the team tracks its performance status, implementation issues, and performance measurement. Each implementation strategy is carried down from Hoshin Implementation Plan. A Hoshin Implementation Review form should be created for each implementation strategy.

For the Six Sigma project to reduce oversize bushing bores, the team has outlined three upcoming steps in their project (see "performance status" in Figure 7.15). These include:

1. Conduct a measurement systems analysis (MSA)
2. Conduct a process failure modes and effects analysis (PFMEA)
3. Conduct design failure modes and effects analysis (DFMEA)

The team has assigned ownership to an individual team member with an expected completion date. A status is also provided by the team as to whether the item is complete, in process, or scheduled.

Hoshin Implementation Review

Core objective:
Implement cost reduction projects

Strategy owner:
Leah

Date:
9/14

Performance status

Conduct MSA	Brian	9/21	Complete
PFMEA	Krista	10/2	Complete
DFMEA	Brooke	11/1	In process

Implementation issues

Mechanic needed on second shift to build gauge stand

Performance measurement

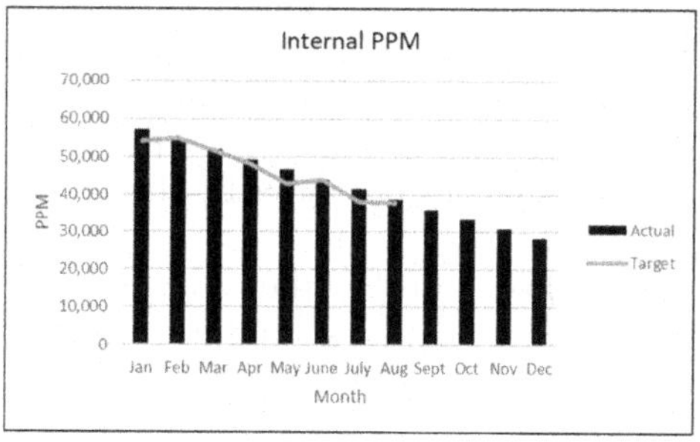

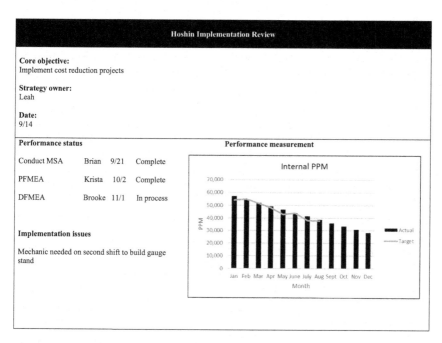

FIGURE 7.15
Carjo's Hoshin Implementation Review.

The next important item is the implementation issues. As part of the measurement systems analysis (see the bottom left of Figure 7.15), the team identified the need for a gage stand. Here, the team also highlighted their trouble with securing a second-shift mechanic to build the gage stand. By presenting this information, the leadership team can secure the necessary resources to make the team successful.

Finally, the team shows their performance trend (on the right-hand side of Figure 7.15). Because the goal is to reduce internal PPM for the bushing bore, the team shows a bar chart of the internal PPM trend.

The Hoshin Implementation reviews should also be posted in the department. The team then can have a stand-up weekly meeting to review the status of the projects. This will help ensure that the projects are on track before the next monthly review with the senior leadership team.

Next Steps for Carjo

The next crucial step for Carjo is to extend the Lean techniques throughout their supply chain. After all, a company is only as good as its suppliers. To develop partnerships and ensure long-term success, the leaders at Carjo should take these methods to their suppliers once they have their internal strategy functioning and engrained into their daily habits. The methodology would use the macro value stream map to show the wastes in the supply chain.

Macro value stream mapping extends beyond the plant-level maps. Macro mapping can be done after creating current and future state maps inside the facility. These maps are created because a large portion of costs consists of purchased materials, and downstream inconsistencies can affect your facility's Leanness. Also, added costs downstream can also negate the cost savings you achieve internally. This can affect whether or not Carjo can grow their sales. The whole picture allows you to identify major asset reconfigurations by showing who does what where and with what tools.

The facility closest to the customer should be mapped first. All of the following information should be collected:

- Frequency
- Distance
- Cost
- Processing time
- Lead time
- Inventories
- Cost per unit
- Daily volume
- Shift data variation
- Frequency variation
- Demand variation

In the ideal state of macro value stream mapping, all activities are located in the exact process sequence.

8

Continuous Improvement Toolkit

The Continuous Improvement Toolkit is an easy reference for what tool to use and when and how to effectively teach the tools to employees who are not necessarily engineers. The implementation of the actual tools will also be taught in this chapter. The Continuous Improvement Toolkit will consist of the following topics that will be taught how to be used with real-life examples:

- 5s
- Affinity diagram
- Analysis of variance (ANOVA)
- Capability analysis
- Cause and effect diagram
- Cellular processing
- Control charts
- Control plan
- Correlation
- CTQ – Critical to Quality
- Defects per million opportunities
- Design for Six Sigma
- DOE – design of experiments
- Failure mode and effect analysis (FMEA)
- Financial justification
- Fishbone diagrams
- Gage R and R
- Graphical analysis
- Histograms
- Hypothesis testing
- Kaizen
- Kanbans
- Kano model

DOI: 10.1201/9780429184123-8

Linear regression
Measurement system analysis
Mood median test
One-point lessons
Pareto charts
Plan Do Check
Poka-yokes
Process capabilities
Process mapping
Project charters
Pull and push flows
Quality function deployment
Root cause analysis
Seven/eight wastes
SIPOC
SMED (single minute exchange of dies)
Spaghetti diagrams
Theory of constraints
Total productive maintenance (TPM)
Value stream mapping
Variation
Visual management
Waste studies

5s

5S is a fundamental tool which serves as the foundation for many tools such as Lean, Six Sigma, total productive maintenance (TPM), and waste management. Future improvements can be made once 5S is implemented. A clean workplace facilitates change where problems naturally stand out. There will also be more room for extra business while ensuring product quality and safety in a well-maintained environment. The goal of 5S is to be able to expose problems that prevent us from being successful in the future.

An unclean environment contains hidden risks for workers and equipment. Advantages of a 5S organization include time management, safety, quality, customer responsiveness, and visual controls. 5S principles will improve work environments where each "S" progresses through implementation.

5S stands for the following:

Sort – Identify and eliminate necessary items and dispose of unneeded materials that do not belong to in an area. This reduces waste, creates a safer work area, opens space, and helps visualize processes. It is important to sort through the entire area. The removal of items should be discussed with all personnel involved. Items that cannot be removed immediately should be tagged for subsequent removal.

Sweep – Clean the area so that it looks like new and clean it continuously. Sweeping prevents an area from getting dirty in the first place and eliminates further cleaning. A clean work place indicates high standards of quality and good process controls. Sweeping should eliminate dirt, build pride in work areas, and build value in equipment.

Straighten – Have a place for everything and everything in its place. Arranging all necessary items is the first step. It shows what items are required and what items are not in place. Straightening aids efficiency; items can be found more quickly and employees travel shorter distances. Items that are used together should be kept together. Labels, floor markings, signs, tape, and shadowed outlines can be used to identify materials. Shared items can be kept at a central location to eliminate purchasing more than needed.

Standardize – Assign responsibilities and due dates to actions while using best practices throughout the workplace. All departments must follow standardized rules to comply with 5S. Items are returned where they belong and routine cleaning eliminates the need for special cleaning projects. Audits to clear up unnecessary items, organize items in designated places, and cleaning are part of the standardized phase. Anything out of place or dirty should be noticed immediately.

Sustain – Establish ways to ensure maintenance of manufacturing or process improvements. Sustaining maintains discipline. Utilizing proper processes will eventually become routine. Training is key to sustaining the effort and involvement of all parties. Management must mandate the commitment to housekeeping for this process to be successful.

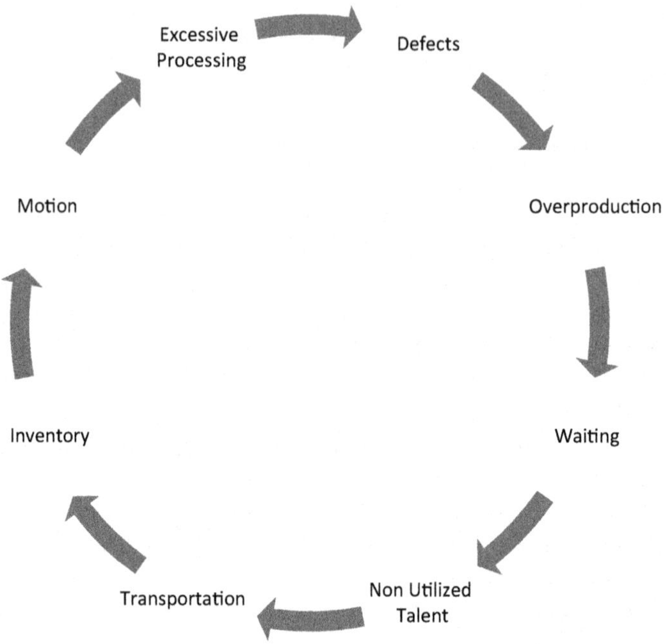

FIGURE 8.1
Eight wastes.

If 5S is not implemented, waste may result. The seven main types of waste are shown in Figure 8.1.

Sort – Segregate what is needed and what is not needed, and what is needed later on. Utilize red tags to determine which items are no longer needed. Discard what is not needed and segregate the needed items by the frequency in which they are used.

Red tags denote items that are not needed. Discard items that are not needed or move them to a storage area. Ensure items that are needed and items that might be needed are differentiated. Move items not needed on weekly and daily basis (Figure 8.2).

These tags must be managed weekly and everyone should be able to have a chance to respond to the red tags. Be critical when looking at these items. Eliminate defective items, store any excess items, and dispose of any obsolete stock.

Sweep – During the sweep phase, items are cleaned up and sometimes deep cleaned. The challenge is to implement regular housekeeping activities so anything dirty is automatically seen

FIGURE 8.2
Red tag.

and is quite obvious. During the sweep phase, undesired dirt or debris is removed and a healthier and safer work environment is promoted. Customers also like the impression of a clean work environment. During this clean phase, inspections are also performed. The items to look for are faulty parts, anything hidden or broken, loose parts or parts in need of calibration, and any items that need to be re-filled. The cleaning must be done on a regular basis. Specific tasks should be assigned to specific personnel. Ensure the proper supplies are close by for the cleaning to take place. A designated time should be set aside per day to do the cleaning as well, this is normally toward the end of the shift.

Straighten – Straightening includes a high level of organization. Arrangement of items should be placed so they are easy to find and be retrieved readily. It should also be obvious if an item is missing. This phase focuses on creating "A place for everything, and everything in its place."

Visual management is a key to this phase to show where, what, and how many. This promotes ready retrieval and ready return. It is also important to remember that people react to colors, shapes, etc., so they should be utilized. Examples can be seen in Figures 8.3–8.5.

Standardize – Communication plays a large role during the standardize phase. Best practices of one area must be communicated along with expectations of other areas. It is important that everyone is aware of their own roles and responsibilities. Training plays a key role in this phase as well so that everyone is trained exactly

FIGURE 8.3
Traffic light visual.

FIGURE 8.4
Before straighten.

the same way. Continuous improvement must be part of the phase as well so that best practices continue to be used. Leaders will set the tone for the best practices to be used. They should gain commitment from employees through their motivation.

Sustain – Celebrating is the best way to motivate employees and ensure they keep on track with their improvements. Recognition is a vital portion of 5S to ensure high standards were met and agreed upon. It is important to sustain results so that the "old" way of doing things doesn't occur. Re-enforcement of the program will ensure its success and convince other workgroups to

FIGURE 8.5
After straighten.

go along the same path. Ensure log books are used for items that were discarded for future use. Pictures should also be taken from the exact same location to show before and after improvements. Action plans for the future must be made for the sustain phase including audits to take place and responsibilities of employees. Continuous improvement again should be utilized to always improve the benefits of 5S (Figure 8.6).

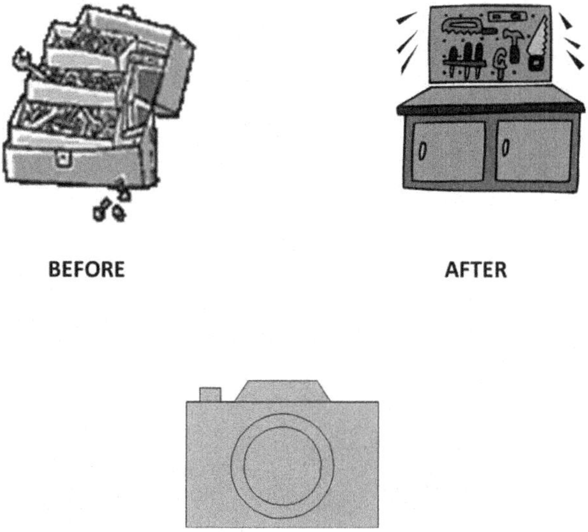

FIGURE 8.6
Before and after picture 5S.

Eight Wastes

The eight wastes are known as the following:

Defects and rework – Defective products or parts

Over production – Producing more than is needed before it is needed

Waiting – Any non-work time waiting for tools, personnel, parts, etc.

Non-utilized talent – Not fully utilizing an employee for a job

Transportation – Wasted efforts used to transport materials, parts, goods, etc., in or out of storage or between processes

Inventory – Excess parts, finished goods, raw materials, etc., that are not being utilized

Movement – Wasted movements to move or store parts or excess walking performed

Excessive processing – Performing more work than is needed (Figure 8.7)

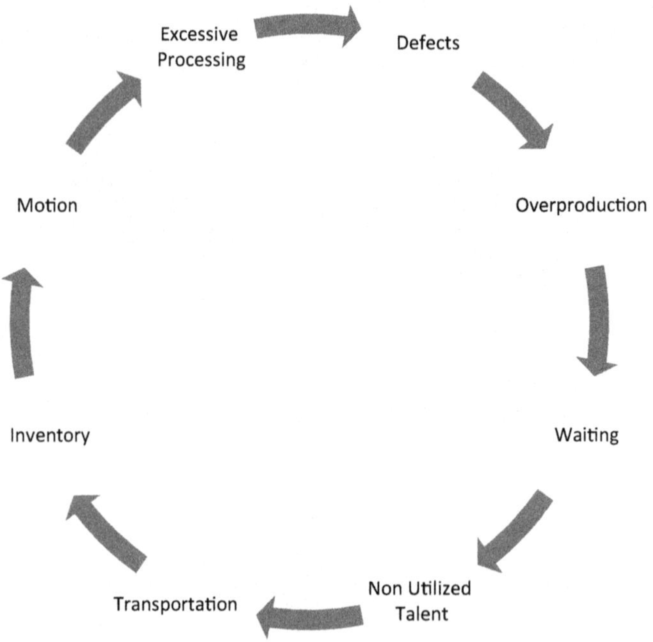

FIGURE 8.7
Eight wastes diagram.

Kaizen

Kaizen is a Japanese word where "Kai" means "Change"
"Zen" means "for the better"
Kaizen – Continuous improvement

The concept includes small incremental improvements that will add up over time. It is important to realize that Kaizen does not necessarily mean a blitz or an event even though an event or blitz often occurs for a Kaizen kick off.

Kaizen is a continuous improvement event with the following:

- Dedicated resources
- Specific goals and deliverables
- Short time frame

It can also be known as

- Rapid continuous improvement event
- Kaizen blitz
- Lean event

Kaizen creates a vision of what our production system and manufacturing techniques should be while carrying out that vision by breaking through the status quo. The basic rules for Kaizen are the following:

- Be open minded for changes
- There are no dumb questions or ideas
- Be positive
- Avoid spending money
- Question current practices and challenge status quo
- Think about how to change versus how not to change
- Go, See, Think, Do
- Have fun (Figure 8.8)

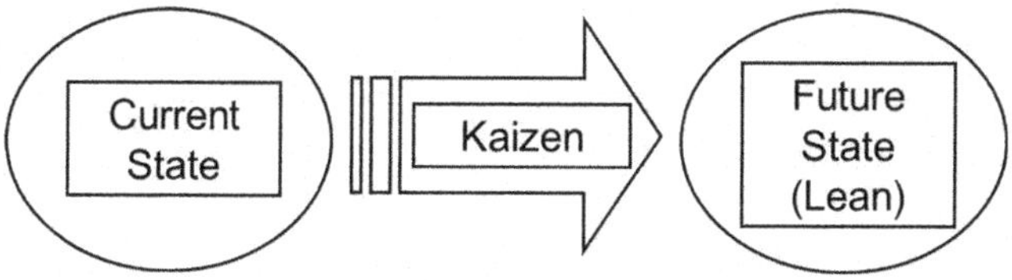

FIGURE 8.8
Current state to future state for Kaizen.

Kaizen drives improvements which lead to a Leaner business operating system.

Fishbone Diagrams

Fishbone diagrams are also known as cause and effect diagrams which are used to understand knowledge about a process or a product. The diagrams were named after Dr. Ishikawa, so sometimes known as an Ishikawa diagram as well. A team would come together to have a structured brainstorming revolving around the process/product. The brainstorming is emphasized in a graphical representation for visual management purposes. It serves as a communication tool. The cause and effect diagram assists in reaching a common understanding of the problem and exposes the potential drivers of the problem. Normally 5Ms are utilized for this brainstorming. The 5Ms are generally Manpower (Personnel is more appropriate), Methods, Materials, Machinery, and Measurements. A sixth element of Environment is frequently used as well. The diagram looks like a fish in that the head will be a box that describes the effect and the body will have bones that include the 5Ms, see Figure 8.9, fishbone diagram. Once each M is defined, a cause and a reason for the cause are identified for the specific M associated within. This type of brainstorming normally uncovers potential issues and results in many action items.

Root Cause Analysis

Special cause variation must be reduced with root cause analysis. This means to not band-aid these problems, but understand **why** these problems happened in the first place.

Root cause analysis is the process of finding and eliminating the cause, which would prevent the problem from returning. Only when the root cause is identified and eliminated can the problem be solved. Root cause analysis uses the D-M-A-I-C philosophy as a guide. The steps to root cause analysis are as follows:

1. Define the problem
2. Map the process
3. Gather data

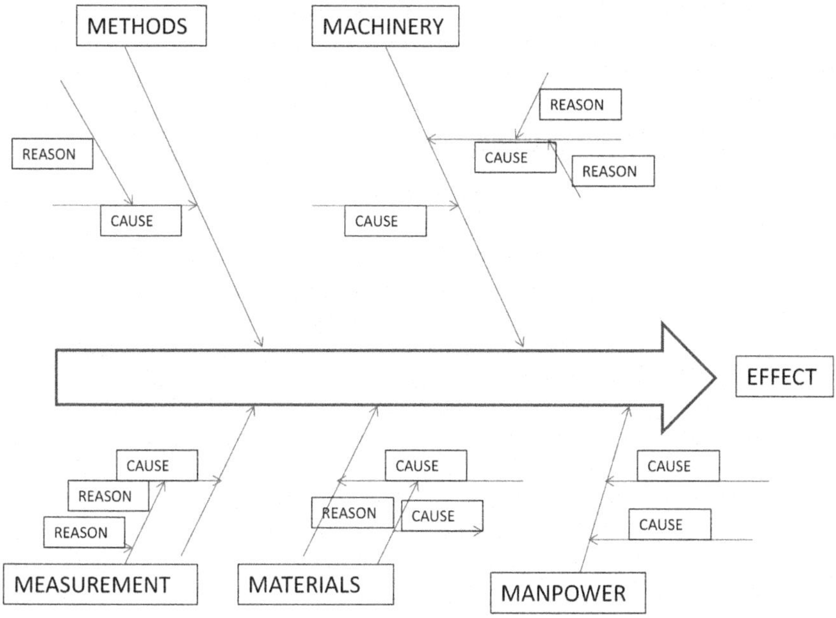

FIGURE 8.9
Fishbone diagram.

4. Seek for root causes through fishbone diagrams/cause and effect diagrams
5. Verify root causes with data
6. Develop solutions and prevention steps including costs and benefits
7. Pilot implementation plans
8. Implement
9. Control utilizing a monitoring plan and process metrics
10. Identify lessons learned

The steps are graphically shown in Figures 8.10–8.15.

Process Mapping

Process maps are graphical representations of steps to a process. The visualization eases the complexity of the process and identifies non-value-added tasks and any key takeaways such as redundancy, excess

Root Cause Analysis		Page 1 of 5
This form is a tool for finding causes of process problems and developing actions for eliminating, preventing, and minimizing future problems.		Date: 1/1/13
Incident Date: 1/1/13	RCA Initiated by: T Agustiady	
Investigator: T Agustiady		
1.Define the problem	Describe the incident. What was defect, how many, how often, etc.	
Line 1 incurs an average of one hour of down time per day on a ten hour shift. Eliminate 50% of mechanical downtime on line by 4/1/13 with Root Cause Analysis		

Step (not all required, depending on problem)		Date completed
D	1 Define the problem	
	2. Map Process (if required)	
M	3. Gather data	
	4. Cause/Effect Analysis (Seeking Root Cause)	
A	5. Verifying root cause with data	
	6. Solutions & Prevention steps development (including cost/benefit)	
I	7. Pilot of implementation	
	8. Implementation	
C	9. Control/Monitoring Plan (including Process Metrics)	
	10. Lessons Learned	

FIGURE 8.10
Root cause analysis.

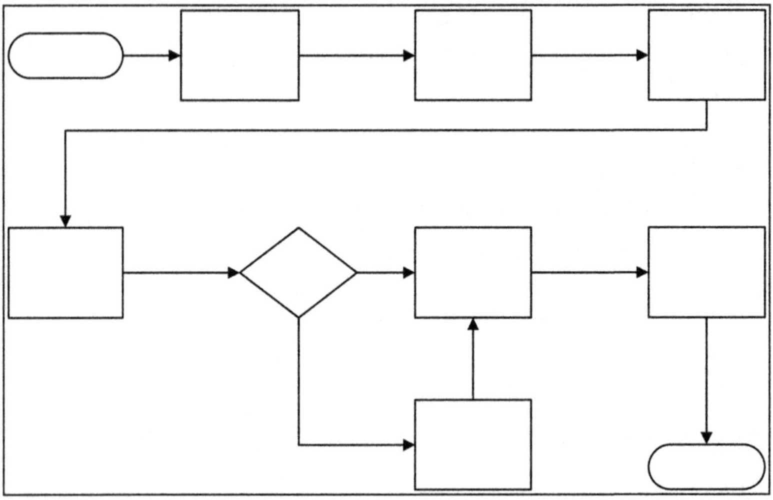

FIGURE 8.11
Process mapping.

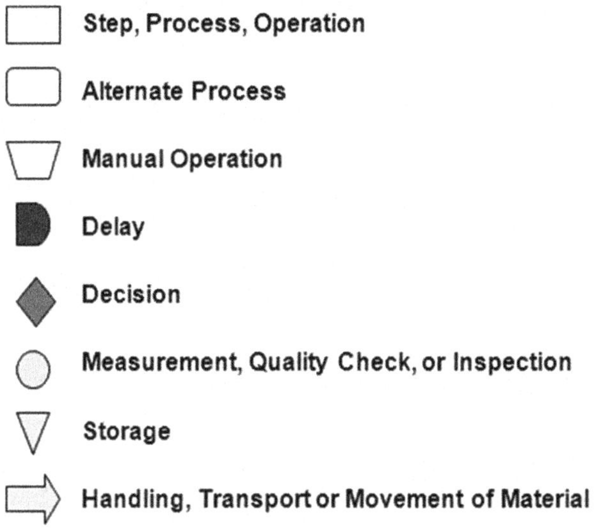

FIGURE 8.12
Process mapping symbols.

inspections. Process maps identify key process input variables known as *x*'s and key process output variables known as *y*'s. Any delays are to be eliminated and decisions are meant to be as efficient as possible. Process maps should be conducted for a specific area at a time due to the complexity of the processes. This way the process maps can be updated frequently by area.

The process mapping symbols are as shown in Figure 8.16.

An example of a process map is shown in Figure 8.17.

Financial Justification

Financial justification is important in order to be able to purchase products, utilize resources, or give return on investments. Cost versus benefits are able to be understood through financial justification. Utilizing these cost benefit tools is essential in order to gain approval from upper management to purchase parts/equipment or put forth resources for a project. Normally a return on investment of less than one year is applicable for the purchase of goods. The identification of costs along with controlling the costs is a financial attribute that should be covered. It should be determined whether goods depreciate or increase in value over time. Benefits

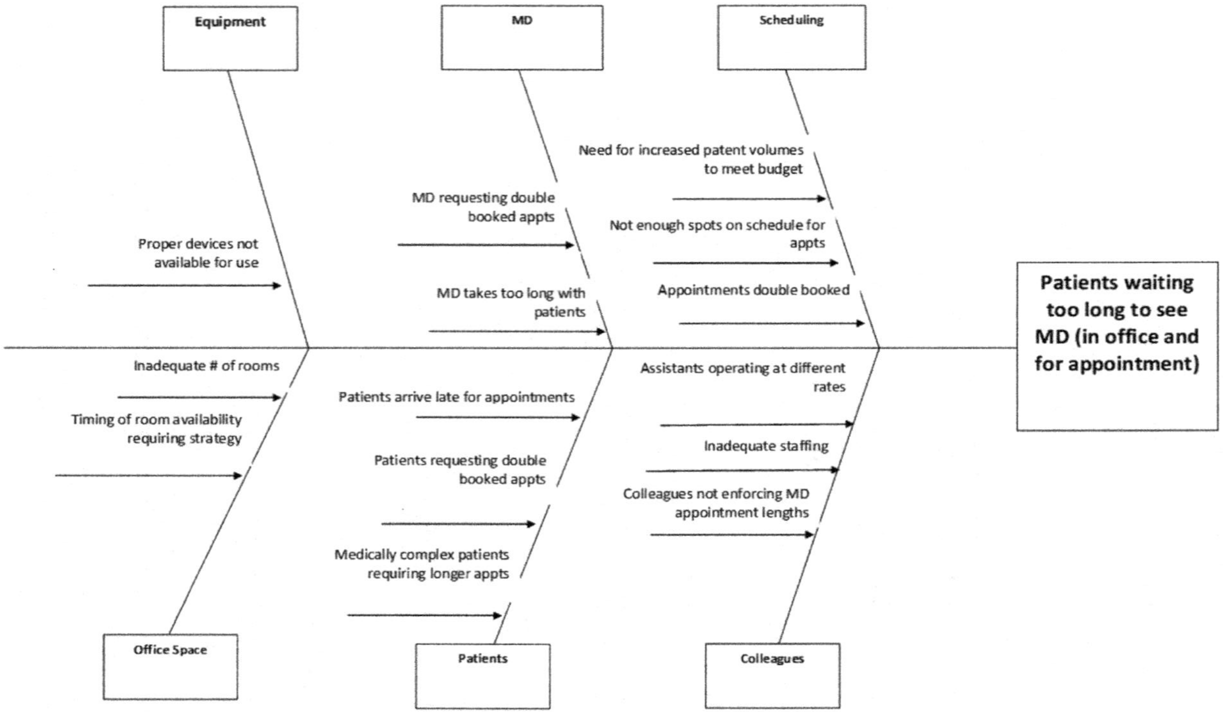

FIGURE 8.13
Fishbone diagram.

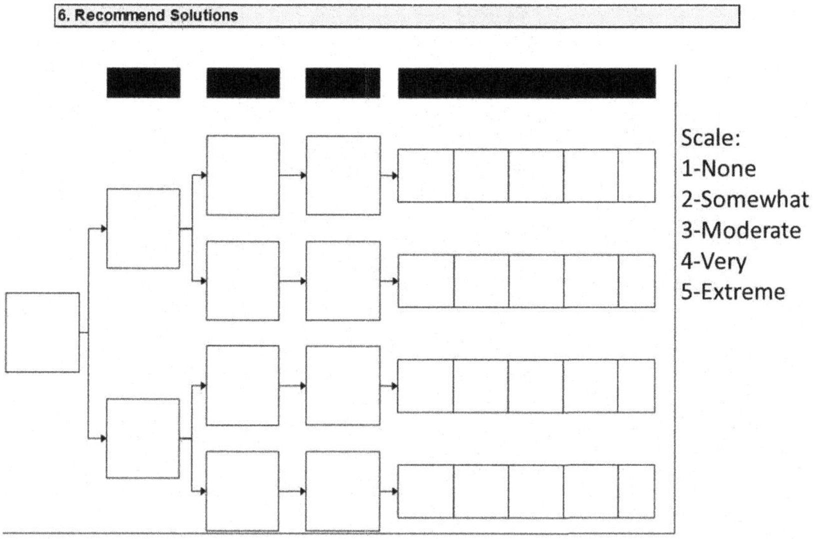

FIGURE 8.14
Recommend solutions.

should always be weighed into consideration beginning with safety and ending with financial justification of a payback. Tangible results from the justification will give approval for the resources or purchase of the goods. Taxes, discounts, and depreciation should be looked at in order to complete financial justifications. Compounding is equivalent to adding the

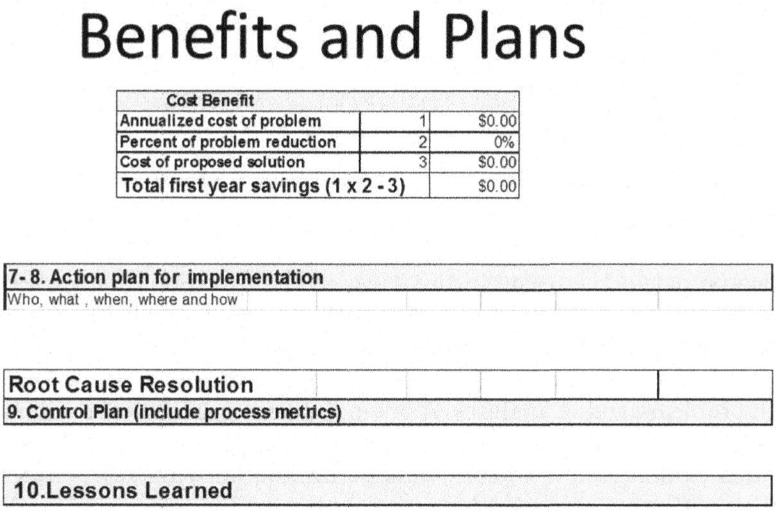

Benefits and Plans

Cost Benefit		
Annualized cost of problem	1	$0.00
Percent of problem reduction	2	0%
Cost of proposed solution	3	$0.00
Total first year savings (1 x 2 - 3)		$0.00

7- 8. Action plan for implementation
Who, what , when, where and how

Root Cause Resolution
9. Control Plan (include process metrics)

10.Lessons Learned

FIGURE 8.15
Benefits and plans with action plans and lessons learned.

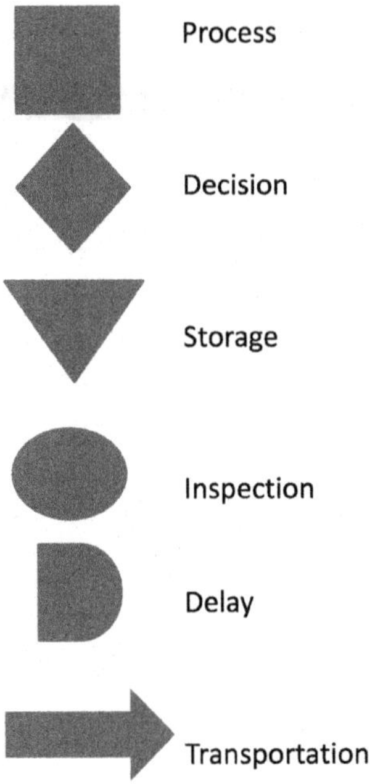

FIGURE 8.16
Process map symbols.

interest to the principal sum, year on year while discounting is equivalent to taking out the interest, backward in time. An example for a cost justification is shown in Figure 8.18 where first-year savings were not met. In this case, it should be calculated what the payback period would be. If the payback is over five years, normally the project is not justified.

One-Point Lessons

A one-point lesson is a simple direct way to give directions that are able to be easily found again and again. A visual documentation helps in giving this lesson so that it is not forgotten. This educational documentation aids in developing employees in their knowledge and skillset while helping with problem-solving analyses. A one-point lesson consists of the general information which includes a title and objective. The lesson will explain the description of how something is performed utilizing key points and visuals. Effective one-point lessons will answer the 5W's and 1H: Who, What, Where, Why, When, and How. An example of a one-point lesson is shown in Figure 8.19.

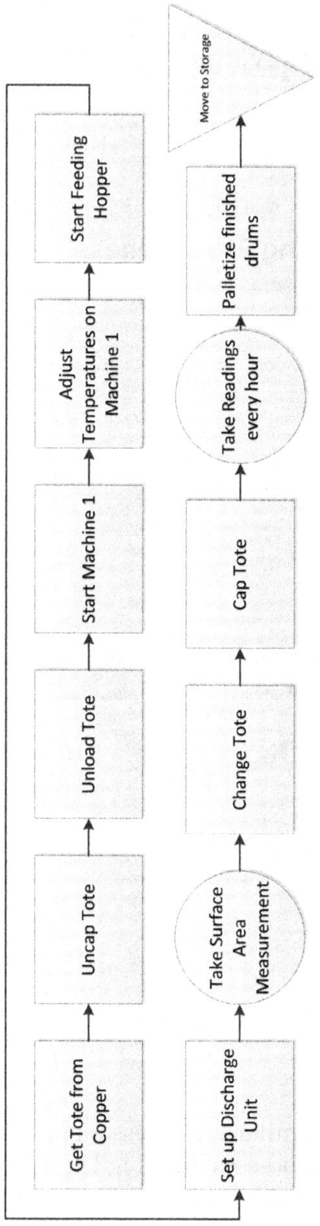

FIGURE 8.17
Process map example.

Cost Benefit		
Annualized cost of problem	1	$100,000
Percent of problem reduction	2	25%
Cost of proposed solution	3	$30,000
Total first year savings (1 x 2 - 3)		-$5,000

FIGURE 8.18
Financial justification example.

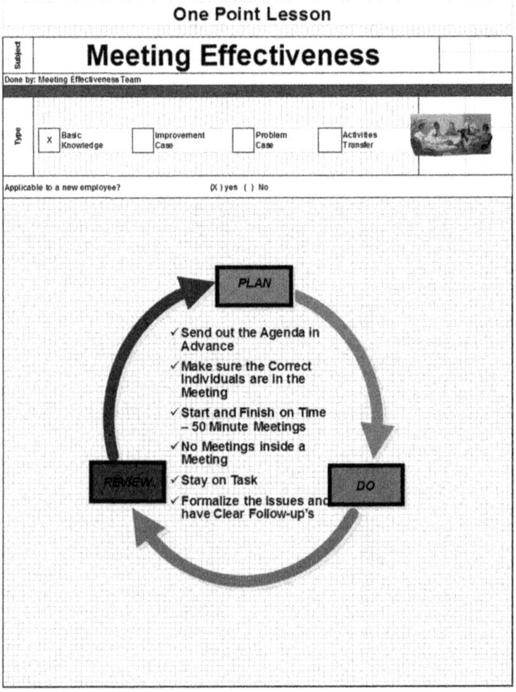

FIGURE 8.19
One-point lesson example.

Value Stream Mapping

Value stream mapping is similar to process mapping yet more advanced because it focuses on the processes utilizing Lean principles according to value. It is important to ensure that a process is considered important if the step adds value. Adding value is also defined as any activity that the customer is willing to pay for. Another note to remember is to not only have a smart and efficient technique, but also only produce goods that the customer is demanding to eliminate excess inventory.

Value stream mapping reveals why excessive waste is introduced. The steps for value stream mapping are as follows:

- Map current-state processes
- Summarize current-state processes
- Map future-state processes
- Summarize future-state processes
- Develop short-term plans to move from current to future state
- Develop long-term plans to move from current to future state
- Implement risks, failures, and processes for transitions
- Map key project owners, key dates, and future dates of future states
- Continue to map new future states when future states are met

All steps in value stream mapping should deliver a product or service to customers.

A value stream involves multiple activities that convert inputs into outputs. All processes follow sequences of starts, lead times, and ends. The first goal is to reduce problematic areas by understanding the main causes for defects. The next step is to reduce lead time. The final step is to find the percentage of value-added time as shown in Figure 8.20. The formula is % VAT = (sum of activity times)/(lead time) × 100. When the sum of activity times equals lead time, the value-added time is 100%. For most processes, % VAT = 5 to 25%. If the sum of activity times equals the lead time, the time value is not acceptable and activity times should be reduced. Lead time is the time from process start to end or time from receipt to delivery. Activity time is the time required to complete one output in an activity. Cycle time is the average time in which an activity yields an output calculated as the activity time divided by the number of outputs produced in one cycle of an activity.

The value stream mapping symbols are shown in Figure 8.20.

An example of an actual value stream map is shown in Figure 8.21.

The main takeaways from the value stream map from the example are the following: C/O (changeover time), shifts, and uptime. The cycle time is able to be calculated with the information showing the value-added activities (Figure 8.22).

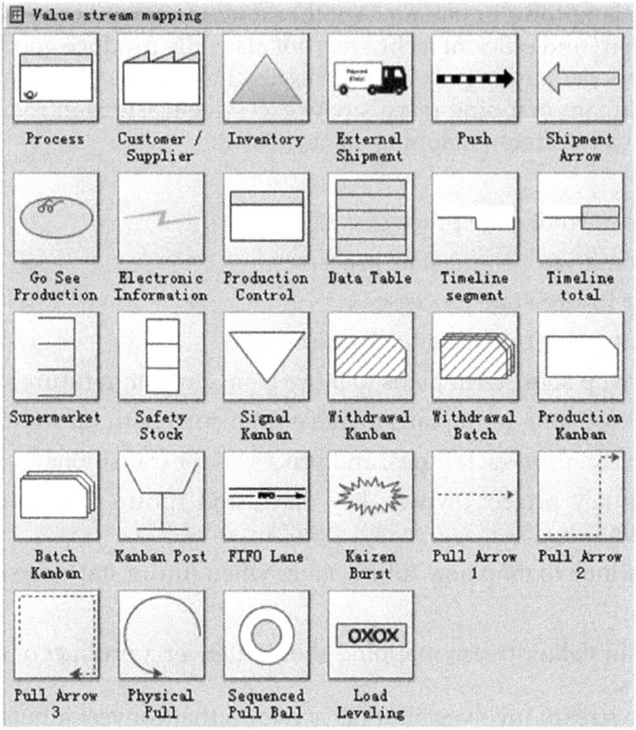

FIGURE 8.20
Value stream mapping basic symbols.

Plan-Do-Check-Act

This is a traditional cycle where processes and conditions are planned out, the planned actions are performed in the Do phase, and finally quality control checks are performed in the Check phase. This method catches mistakes and also provides feedback during the Check phase. The checks in this place also account for 100% inspection, therefore all parts or processes are looked upon indicating no defects (Figure 8.23).

There are three main types of checks or inspections that are popular:

- Judgment inspections
- Informative inspections
- Source inspections

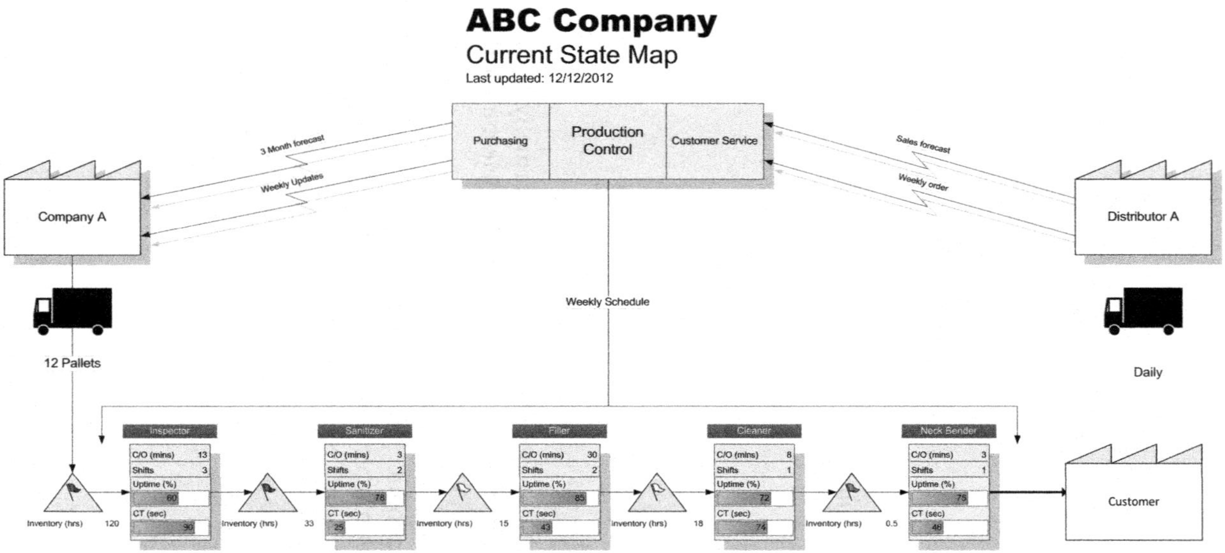

FIGURE 8.21
Value stream map example.

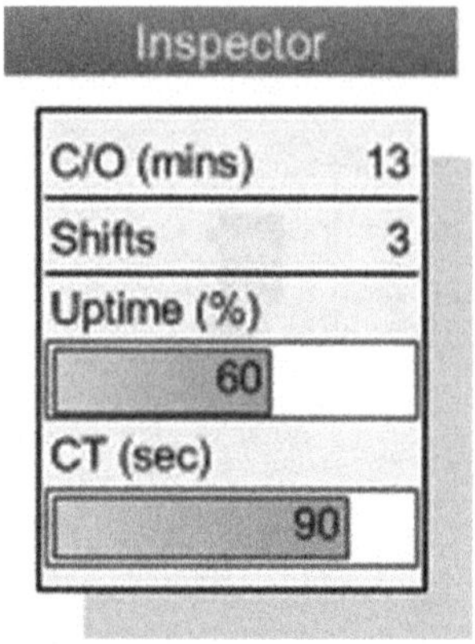

FIGURE 8.22
Value stream map cycle time.

FIGURE 8.23
Plan-do-check-act illustration.

Judgment inspections are those that are done normally by humans based on what their expectations are. They find the defect after the defect has already occurred.

Informative inspections are based on statistical quality control (SQC), checks on each product, and self-checks. These inspections help reduce defects, but not eliminate them completely. Finally, the source inspections are the inspections that reduce the defects completely. Source inspections discover the mistakes before processing and then provide feedback and corrective actions so that the process has zero defects. The source inspections require 100% inspection. The feedback loop is also very quick so that there is minimal waiting time.

Poka-yokes

A poka-yoke is a Japanese term that means "mistake proofing." All inadvertent defects can be prevented from happening or prevented from being passed along.

Poka-yokes use two approaches:

- Control systems
- Warning systems

Control systems stop the equipment when a defect or unexpected event occurs. This prevents the next step in the process to occur so that the complete process is not performed. Warning systems signal operators to stop the process or address the issue at the actual time. Obviously, the first of the two prevents all defects and has a more zero quality control (ZQC) methodology because an operator could be distracted or not have time to address the problem. Control systems often also use lights or sounds to bring attention to the problem, and that way the feedback loop again is very minimal.

The methods for using poka-yoke systems are as follows:

- Contact methods
- Fixed-value methods
- Motion-step methods

Contact methods are simple methods that detect whether or not products are making physical or energy contact with a sensing device. Some of these are commonly known as limit switches where the switches are connected to cylinders and pressed in when the product is in place. If a screw is left out, the product does not release to the next process. Other examples of contact methods are guide pins.

Fixed-value methods are normally associated with a particular number of parts to be attached to a produce or a fixed number of repeated operations occurring at a particular process. Fixed-value methods utilized devices as counting mechanisms. The fixed-value methods may also use limit switches or different types of measurement techniques.

Finally, the motion-step method senses if a motion or step in the process has occurred in a particular amount of time. It also detects sequencing by utilizing tools such as photoelectric switches, timers, or barcode readers.

The conclusion of poka-yokes is to use the methodology as mistake proofing for ZQC to eliminate all defects, not just some. The types of poka-yokes do not have to be complex or expensive, just well thought out to prevent human mistakes or accidents.

The poka-yoke discussion stems into the correct location discussion. This technique places design and production operations in the correct order to satisfy customer demand. The concept is to increase throughput of machines ensuring the production is performed at the proper time and place. Centralization of areas helps final assemblers, but the most

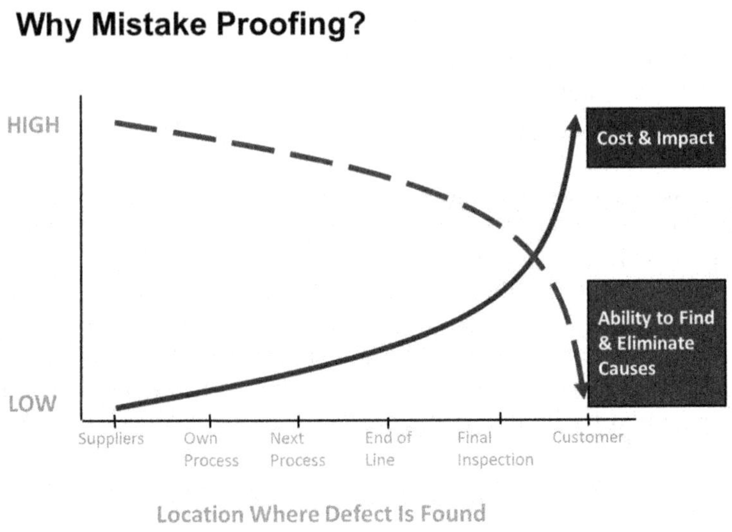

FIGURE 8.24
Why mistake proofing?

common practice to be effective is to unearth an effective flow. U-shaped flows normally prevent bottlenecks. Value stream mapping is a key component during this time in order to establish that all steps occurring are adding value. A reminder for value-added activities: any activity that the customer is willing to pay for. Another note to remember is to not only have a smart and efficient technique, but also only produce goods that the customer is demanding to eliminate excess inventory (Figure 8.24).

The advantages of mistake proofing are the following:

- No formal training required
- Relieves operators from repetitive tasks
- Promotes creativity and value-added activities
- 100% inspection internal to the operation
- Immediate action when problems arise
- Eliminates need for many external inspection operations
- Results in defect-free work

Errors are the cause.

- An error occurs when the conditions for successful processing are either incorrect or absent

Defects are the result.

Defects are prevented if

- Errors are prevented from happening
- Errors are discovered and eliminated (Figure 8.25)

How to mistake proof:

- Adapt the right attitude
- Select a process to mistake proof
- Select a defect to eliminate
- Determine the source of the defect
- Identify countermeasures
- Develop multiple solutions
- Implement the best solution
- Document the solution

Which processes should be mistake proofed?

- High error potential
- Complex processes
- Routine "boring" processes
- High failure history
- Critical process characteristic
- High scores in FMEA
- Use problem history

Error		Defect
Not setting the timer properly on your toaster		Burnt toast
Placing the "originals" in your copier "Face Up"		Blank pages
Running out of ink on your date stamp		No date stamp on paper

FIGURE 8.25
Mistake proofing examples.

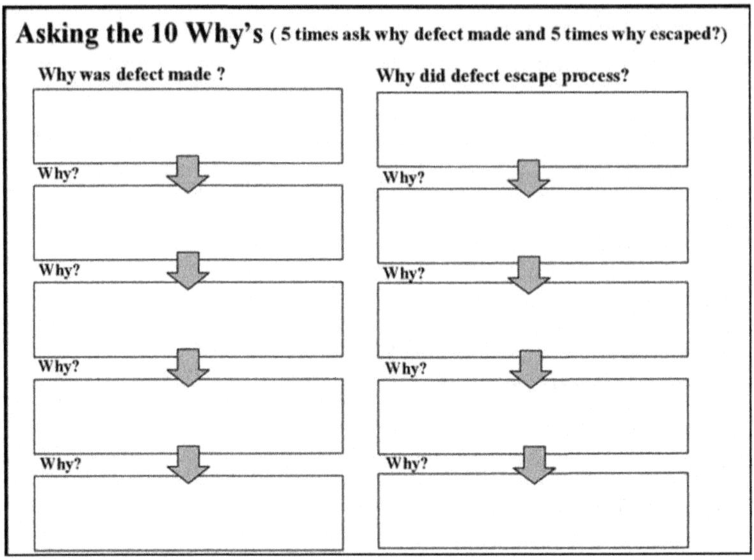

FIGURE 8.26
10 why's for mistake proofing.

- Pareto charts, etc.
- Isolate the specific defect
- Do not be too general
- Do not combine defects
- Make a decision on which defect to work on based on
- Severity
- Frequency
- Ease of solving
- Annoyance factor (Figure 8.26)

Successive checking – positives and negatives: Figure 8.27.

Kanbans

Kanbans are communication signals that control inventory levels while ensuring even and controlled production flow. Kanbans signal times to start, times to change setups, and times to supply parts. Kanbans work only if monitored consistently. Kanbans are used in Lean production

Successive Checks	Self Check Detection in Station	Mistake Proof (Poka-Yoke)
Associates check work of previous associate	Associates check own work before passing to the next associate	Automatic check and prevention of defect
Plus: Generally effective in catching defects	Plus: Instant correction possible and more palatable than supervisor check or peer check	Plus: 100% inspection usually with no extra time expense with the benefit of instant correction
Corrective action can only occur after defect is made	Associate may compromise quality or forget to perform self-check	None

FIGURE 8.27
Different methods of checks.

to ensure that flow is pulled by the next step and a visual indicator signals the needs for inventory or activity. These steps enable production to respond directly to customer need without producing excess inventory or requiring further work. Quality should also soar because production is based on customer needs and not production per minute or per hour (Figure 8.28).

Pull and Push Flows

Pull is the simple practice of not producing a product upstream until the customer downstream needs it. The philosophy sounds easy, but its management is complicated. Normally large batches of general inventory are in stock, but ensuring specialized and costly inventories for specific customers is far more difficult. Understanding how to be versatile and utilizing comanufacturers for these special circumstances are keys to Lean

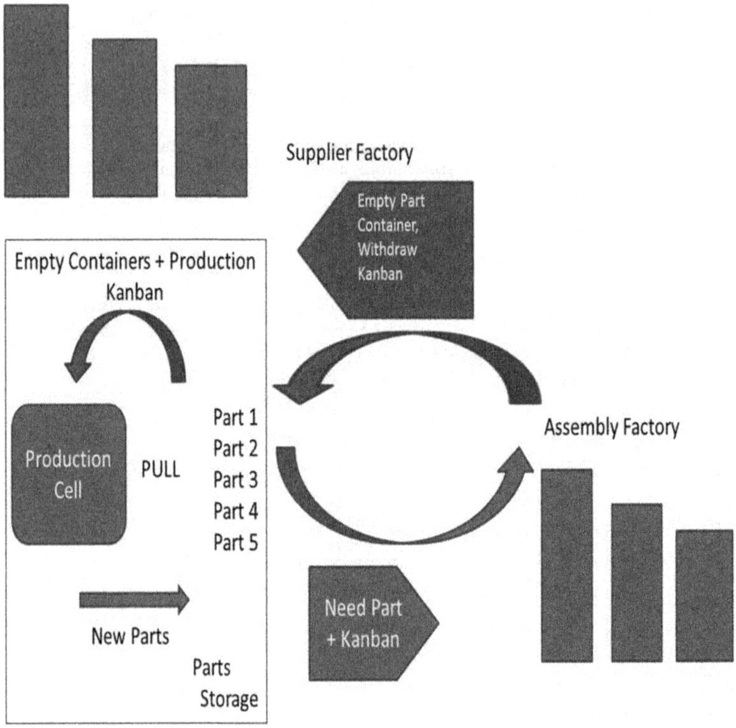

FIGURE 8.28
Kanban example.

production utilizing a pull system. Customers for high-end or specialized products must understand the lead times required.

Push systems are not accurate because they normally involve production schedules based on projected customer demands. Changeovers should be kept to a minimum to produce versatile products instead of large quantities of a special product. Small "stores" of parts between operations are created to reduce inventory. When a process customer uses inventory, it should be replenished. If inventory is not needed, it is not replenished. Lean production for pull requires 90% machine use or availability and limits downtime for maintenance and changeovers to 10% of machine time. Kanbans are used in Lean production to ensure that flow is pulled by the next step and a visual indicator signals the needs for inventory or activity. These steps enable production to respond directly to customer need without producing excess inventory or requiring further work. Quality should also soar because production is based on customer needs and not production per minute or per hour.

Visual Management

Visual boards show team successes and give communication without every team member being present. A visual management system normally consists of some type of a large white board that has key points that are important to each team member. Common visual boards consist of the following points:

- Each person writes down three key items they are working on and what their hold up is
- Recognition of a key associate
- Capital spending box
- Reminders
- Calendar reference
- Meeting time
- Overdue action items

It is important to meet once a week at the same time to review the visual management board. Each person is accountable for their box and should fill it out before the meeting.

Utilizing different colors for certain topics such as pink for important items helps people to visually separate items. Suggestions from the visual management boards should be taken, because the board should always be changing and improving. It is also important to remember that people react to colors, shapes, etc., so they should be utilized (Figure 8.29).

Any types of measurements that can be seen visually are also key indicators if a process is performing well or not. A visual technique for this is shown in a simple measuring cup in Figure 8.30.

The purpose of these visual controls is to show what's right, what's wrong, what's complete, where the path going forward should be, what step to take next, who to see, and common standard operating procedures.

FIGURE 8.29
Visual management technique.

FIGURE 8.30
Visual management technique.

How does a Visual Board help create Lean Engineering?

Stability

- Implementation of Lean will lead to stability of technical and work processes
    - Weekly Visual Board Meetings

Work efficiency and effectiveness

- Stability will lead to higher work efficiency and effectiveness
    - Reminders for training, meetings, upcoming events
- Higher work efficiency and effectiveness will free up resources
    - Resolves project hold ups for the group; visual group work load balance for manager

Continuous Improvement

- Resources can be used for continuous improvement for leading the sites to world class chemical sites
    - The visual board is always changing based on the groups needs to get their jobs done efficiently

FIGURE 8.31
Visual management board and Lean engineering.

These techniques tell us at a glance how things are going and point out imperfections or abnormal conditions instantaneously. Visual techniques will enhance performance to the next level improving office and factory performances (Figure 8.31).

Cellular Processing

Centrally locating cells within all machines and work stations are important so that the final product can be completed without major delays of transportation. The premise is to reduce inventory, lead time, material handling, intermediate storages, and improve communication between operations. One operator should be able to run multiple processes by utilizing centralized cellular management.

Processes and equipment that are alike should be placed together. Takt times can be calculated in order to increase uptime and standardize process times. Utilizing a U-shaped layout ensures the personnel is located centrally with ease of access to machinery and workstations. The quality of inspection is increased utilizing this layout as well. The capacity of the cell is increased by only making to customer orders by calculating the demand needed for all items. Cycle times should be calculated for all parts of machinery and processing times, operator load times, and setup and changeover times. If the capacity of one operation is higher than the demand for the product, the machinery should be able to sit idle. The bottleneck in this case should be identified, and the downtime should be eliminated for it. The reduction of setup times and changeover times is the ultimate goal.

There are five main steps in making this cellular layout:

- Group similar products
- Measure demands through takt time
- Review work sequences
- Combine work in balanced processes
- Design the cellular layout

U-shaped layouts are generally the most efficient since they provide the shortest distance for time traveled. Work also enters and exits at the same vicinity while communication is increased and flexibility can be completed by different operators that are able to be cross functional. A modified circle is also another option which has the same premises of the U.

Mapping the Flow with Spaghetti Diagrams

In order to map the current flow to reduce time, an analysis of movement should be performed. Simple ways to map flow is by utilizing a spaghetti diagram. Flows should be continuous, and visually illustrating them is an easy way to determine travel distances. A spaghetti diagram can be completed by simply drawing out the layout of equipment and then mapping a person's walking paths through the production day (Figure 8.32).

Virtually all steps should be mapped out with lines showing where the person travels. After the travel time is mapped out, it can be seen where the bottlenecks lie. The step after this is to perform the future state of the spaghetti diagram (Figure 8.33).

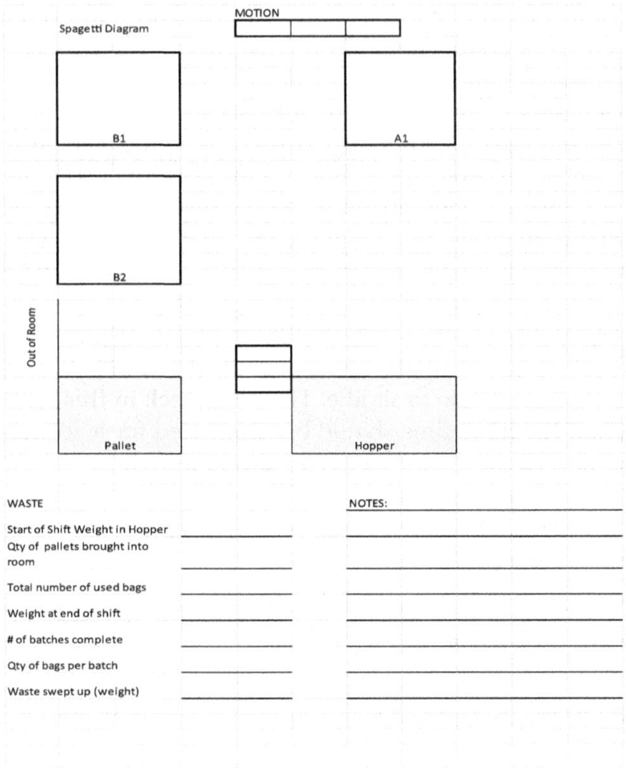

FIGURE 8.32
Spaghetti diagram layout.

The steps for mapping a spaghetti diagram are the following:

- Step 1 – Create a map of the work area layout.
- Step 2 – Observe the current work flow and draw the actual work path from the very beginning of work to the end when products exiting the work area.
- Step 3 – Analyze the spaghetti chart and identify improvement opportunities.

Histograms

A histogram plots the frequency of values grouped together as a bar graph. Histograms are handy for determining location, spread, and shape. Outliers can easily be identified. The height equals the frequency and the width equals a range of values. A histogram with a bell-shaped curve is normal (Figure 8.34).

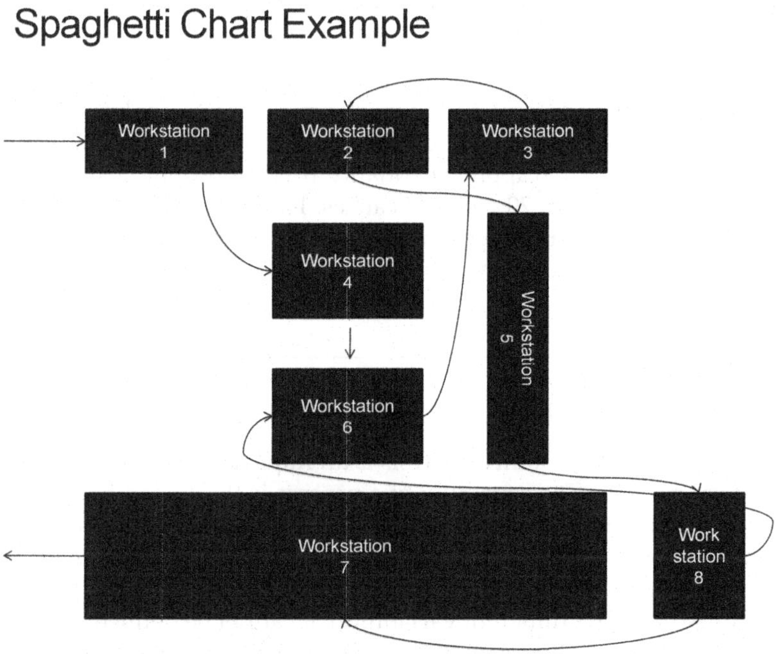

FIGURE 8.33
Spaghetti diagram.

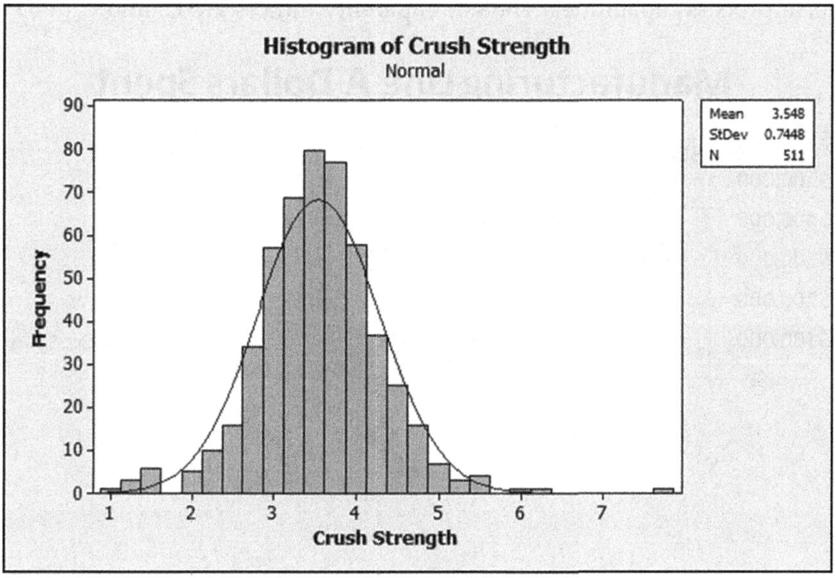

FIGURE 8.34
Histogram example.

Pareto Charts

Key projects can be determined by performing Pareto analysis. This statistical tool implies that by doing 20% of the work, 80% of the advantage can be generated. When applied to quality, this philosophy states that 80% of problems stem from 20% of key causes. Pareto analyses are guides to prioritizing and determining key opportunities (Figure 8.35).

Capability Analysis

Industrial process capability analysis is an important aspect of managing industrial projects. The capability of a process is the spread that contains most of the values of the process distribution. It is very important to note that capability is defined in terms of distribution. Therefore, capability can only be defined for a process that is stable (has distribution) with common cause variation (inherent variability). It cannot be defined for an out-of-control process (no distribution) with variation arising from specific causes (total variability). The key need for capability analysis is to ensure the process is meeting the requirements. Capability analysis can be done with both attribute and variable data. They measure the short- and long-term process capabilities. The key capability indices are C_p and C_{pk} or P_p

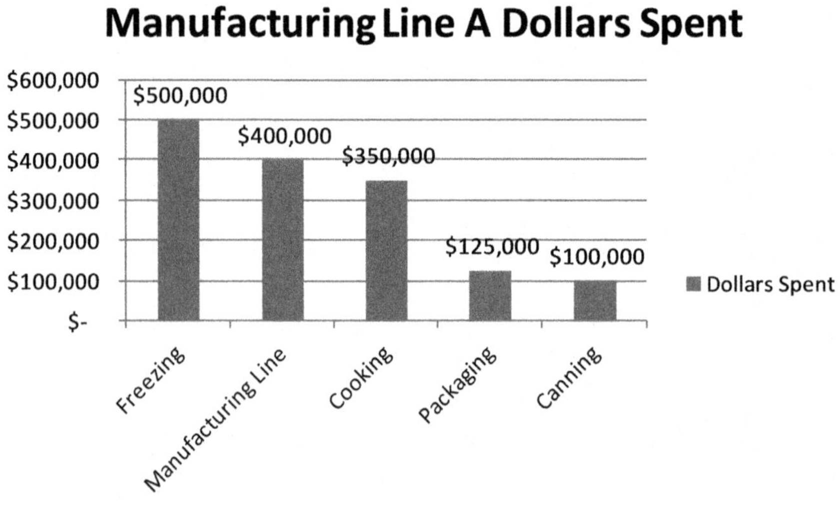

FIGURE 8.35
Pareto chart.

$$C_p = \frac{USL - LSL}{6\hat{\sigma}}$$
$$C_pU = \frac{USL - \overline{X}}{3\hat{\sigma}}$$
$$C_pL = \frac{\overline{X} - LSL}{3\hat{\sigma}}$$
$$C_{pk} = Min(C_pU,\ C_pL)$$

FIGURE 8.36
Capability indices.

and P_{pk} for long-term capabilities. The formulas for calculating the capability indices are provided in Figure 8.36.

The value of C_{pk} should be greater than 1.5 in order to be of a "good" capability. C_p and C_{pk} use a pooled estimate of the standard deviation (SD), whereas P_p and P_{pk} use the long-term estimate of the standard deviation. C_{pk} is what the process is capable of doing if there is no subgroup variability and P_{pk} is the actual process performance. Normally, C_{pk} is smaller than P_{pk} since P_{pk} represents both within subgroup and between subgroup variability. C_{pk} only represents between subgroup variability. The main steps to a capability study consist of the following:

1. Set up the process to the best parameters and identify key process input variables
2. Identify subgroups
3. Run the product over a short time span to minimize the impact of special cause variation
4. Observe the process and take notes throughout
5. Measure and identify key process output variables
6. Run capability analyses to review normality, statistical process control, and histograms
7. Run capability analyses for short-term and total standard deviations
8. Identify mean shifts and variation
9. Estimate long-term capability
10. Develop action plans based on the data

Setting short-term and long-term goals based on capability analyses will result in successful action plans based on real-time data.

The goal of capability studies is to do the following:

Move the P_{pk} to P_p to center the process

Move the P_p to C_{pk} to reduce the variation

Move the C_p to C_{pk} to reduce random variation

A Six Sigma process has a C_p of 2.00 and a P_{pk} of 1.5.

Control Charts

Control charts may be used to track project performance before deciding what control actions are needed. Control limits are incorporated into the charts to indicate when control actions should be taken. Multiple control limits may be used to determine various levels of control points. Control charts may be developed for costs, scheduling, resource utilization, performance, and other criteria. Control charts are extensively used in quality control work to identify when a system has gone out of control. The same principle is used to control the quality of work sampling studies. The 3σ limit is normally used in work sampling to set the upper and lower limits of control. First the value of p is plotted as the center line of a p-chart. The variability of p is then found for the control limits. Two of the most commonly used control charts in industry are X-bar charts and range (R) charts. The type of chart depends on the kind of data collected: variable data or attribute data. The success of quality improvement depends on (1) the quality of data available and (2) the effectiveness of the techniques used for analyzing the data. The charts generated by both types of data are as follows:

Variable data: control charts for individual data elements (X)

Moving range chart (MR-chart)

Average chart (X-chart)

Range chart (R-chart)

Median chart

Standard deviation chart (σ-chart)

Cumulative sum chart (CUSUM)

Exponentially weighted moving average (EWMA)

Attribute data:

Proportion or fraction defective chart (p-chart); subgroup sample size can vary.

Percent defective chart (100p-chart); subgroup sample size can vary.

Number defective chart (np-chart); subgroup sample size is constant.

Number defective (c-chart); subgroup sample size = 1.

Defective per inspection unit (u-chart); subgroup sample size can vary. The statistical theory useful to generate control limits is the same for all the charts except the EWMA and CUSUM charts.

X-Bar and Range Charts

The R-chart is a time plot useful for monitoring short-term process variations. The X-bar chart monitors long-term variations where the likelihood of special causes is greater over time. Both charts utilize control lines called upper and lower control limits and central lines; both types of lines are calculated from process measurements. They are not specification limits or percentages of the specifications or other arbitrary lines based on experience. They represent what a process is capable of doing when only common cause variation exists, in which case the data will continue to fall in a random fashion within the control limits and the process is in a state of statistical control. However, if a special cause acts on a process, one or more data points will be outside the control limits and the process will no longer be in a state of statistical control.

Calculation of Control Limits

- Range (R)

  This is the difference between the highest and lowest observations:

  $R = X_{\text{highest}} - X_{\text{lowest}}$

- Center lines

  Calculate $\bar{X}$ and $\bar{R}$

$$\bar{X} = \frac{\sum X_i}{m}$$

$$\bar{R} = \frac{\sum R_i}{m}$$

where

$\bar{X}$ = overall process average

$\bar{R}$ = average range

m = total number of subgroups

n = within subgroup sample size

- Control limits based on R-chart

  $UCL_R = D_4 \bar{R}$

  $LCL_R = D_3 \bar{R}$
- Estimate of process variation

$$\hat{\sigma} = \frac{\bar{R}}{d_2}$$

- Control limits based on $\bar{X}$ -chart

  Calculate the upper and lower control limits for the process average:

$$UCL = \bar{X} + A_2\bar{R}$$

$$LCL = \bar{X} - A_2\bar{R}$$

The values of d_2, A_2, D_3, and D_4 are for different values of n. These constants are used for developing variable control charts.

Plotting Control Charts for Range and Average Charts

- Plot the range chart (R-chart) first.
- If R-chart is in control, then plot X-bar chart.
- If R-chart is not in control, identify, and eliminate special causes, then delete points that are due to special causes, and re-compute the control limits for the range chart. If process is in control, then plot X-bar chart.
- Check to see if X-bar chart is in control, if not search for special causes and eliminate them permanently.
- Remember to perform the eight trend tests.

Plotting Control Charts for Moving Range and Individual Control Charts

- Plot the moving range chart (MR-chart) first.
- If MR-chart is in control, then plot the individual chart (X).

- If MR-chart is not in control, identify, and eliminate special causes, then delete special causes points, and re-compute the control limits for the moving range chart. If MR-chart is in control, then plot the individual chart.
- Check to see if individual chart is in control, if not search for special causes from out-of-control points.
- Perform the eight trend tests.

Defects per Million Opportunities

Six Sigma provides tools to improve the capabilities of business processes while reducing variation. It leads to defect reduction and improved profits and quality. Six Sigma is a universal scale that compares business processes based on their limits to meet specific quality limits. The system measures defects per million opportunities (DPMOs). The Six Sigma name is based on a limit of 3.4 defects per million opportunities.

Figure 8.37 shows a normal distribution that underlies the statistical assumptions of the Six Sigma model. The Greek letter σ (sigma) marks the distance on the horizontal axis between the mean μ and the curve inflection point. The greater this distance, the greater is the spread of values encountered. The figure shows a mean of 0 and standard deviation of 1, that is, $\mu = 0$ and $\sigma = 1$. The plot also illustrates the areas under the normal curve within

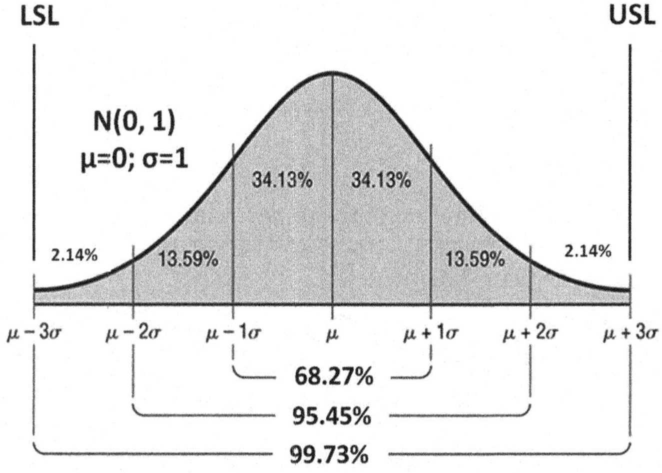

FIGURE 8.37
Areas under normal curve.

TABLE 8.1

Defects per Million Opportunities and Sigma Levels

Sigma Spread	DPMO	Percent Defective (%)	Percent Yield (%)	Short-Term C_{pk}	Long-Term C_{pk}
1	691,462.00	69	31	0.33	−0.17
2	308,538.00	31	69	0.67	0.17
3	66,807.00	7	93.3	1	0.5
4	6,210.00	0.62	99.38	1.33	0.83
5	233.00	0.02	99.98	1.67	1.17
6	**3.40**	**0**	**100**	**2**	**1.5**

different ranges around the mean. The upper and lower specification limits (USL and LSL) are $\pm 3\sigma$ from the mean or within a Six Sigma spread.

Because of the properties of the normal distribution, values lying as far away as $\pm 6\sigma$ from the mean are rare because most data points (99.73%) are within $\pm 3\sigma$ from the mean except for processes that are seriously out of control.

Six Sigma allows no more than 3.4 defects per million parts manufactured or 3.4 errors per million activities in a service operation. To appreciate the effect of Six Sigma, consider a process that is 99% perfect (10,000 defects per million parts). Six Sigma requires the process to be 99.99966% perfect to produce only 3.4 defects per million, that is 3.4/1,000,000 = 0.0000034 = 0.00034%. That means that the area under the normal curve within $\pm 6\sigma$ is 99.99966% with a defect area of 0.00034%. Six Sigma pushes the limit of perfection! Table 8.1 depicts long-term DPMO values that correspond to short-term sigma levels.

Project Charters

A project charter is a definition of the project that includes the following:

- Provides problem statement
- Overview of scope, participants, goals, and requirements
- Provides authorization of a new project
- Identifies roles and responsibilities

Once the project charter is approved, it should not be changed.

A project charter begins with the project name, the department of focus, the focus area, and the product or process.

An example of the table of contents for a project charter is shown in Figure 8.38.

A project charter serves as the focus point throughout the project to ensure the project is on track and the proper people are participating and being held accountable.

The importance of a project charter with respect to sustainability is the living document to educate and give governance for a new project.

TABLE OF CONTENTS

FIGURE 8.38
Project charter example.

Sustainability needs to utilize a great deal of education while giving goals and objectives. A project charter will serve as this living document for organizations with specified approaches.

SIPOC

The SIPOC identifies

1. Major tasks and activities
2. The boundaries of the process
3. The process outputs
4. Who receives the outputs (the customers)
5. What the customer requires of the outputs
6. The process inputs
7. Who supplies the inputs (suppliers)
8. What the process requires of the inputs
9. The best metrics to measure

SIPOC stands for the following and is shown in Figure 8.39.

Supplier – Know and work with your supplier while working with your supplier to help them improve.

Input – Strive to continually improve the inputs by trying to do the right thing the first time.

Process – Describe the process at a high level, but with enough detail to demonstrate to an executive or manager. Understand the process fully by knowing it 100%. Eliminate any mistakes by implementing a poka-yoke.

Output – Strive to continually improve the outputs by utilizing metrics.

Customer – Keep the customer's requirements in sight by remembering they are the most important aspect of the project. The customer makes the specifications, keep the Critical to Quality (CTQs) of the customer in mind.

FIGURE 8.39
SIPOC.

SIPOC steps:

1. Gain top-level view of the process
2. Identify the process in simple terms
3. Identify external inputs such as raw materials and employees
4. Identify the customer requirements also known as outputs
5. Make sure to include all value-added and non-value-added steps
6. Include both process and product output variables

SIPOC implies that the process is understood and helps easily identify opportunities for improvement.

A SIPOC is important in concepts of sustainability because it helps develops a solution for development. Normally the process is mapped out as a high level first.

An example of a SIPOC for assembling a part is provided in Figure 8.40.

The important part of a SIPOC is to look at the details of the current state and see what improvements can be made for future states. Adding specifications for any of the inputs can identify gaps in the process. Benchmarking one process to another will also identify gaps.

Suppliers	Inputs	Input Specification	Processes	Gap	Outputs	Customers
Raw Bone	Thread	1.0 - 1.1	Materials come from warehouse to Xline		Raw Material	Wally
Pet Doo	Material	2.0 - 2.9	Material fabricated by machine X		Fabricated Material	Tget
Cryer	Bone	4.1 - 4.9	Toy sent to machine Y		1/2 processed toy	Kroman
Whiner	Cushion	.8 - 1.0	Material put together by machine Y		Toy complete for inspection	Giant Store
Happy Pup	Squeaker	5.5 - 6.5	QC Process	**Manual Process**	Toy complete for inspection	Pet Peeps
	Rope	10.1 - 10.9	Re-work		Toy complete for inspection	
	Treat	2.4 - 2.10	Decorate product		1/2 processed toy	
			Package		Fully processed toy	
			Ship		Toy ready for customer	

FIGURE 8.40
SIPOC example.

Kano Model

The Kano model was developed by Noriaki Kano in the 1980s. The Kano model is a graphical tool that further categorizes Voice of the Customer (VOC) and CTQs into three distinct groups:

- Must-haves
- Performance
- Delighters

The Kano helps identify CTQs that add incremental value versus those that are simply requirements where having more is not necessarily better.

The Kano model engages customers by understanding the product attributes, which are most likely important to customers. The purpose of the tool is to support product specifications which are made by the customer and promote discussion while engaging team members. The model differentiates the features of products rather than customer needs by understanding necessities and items that are not required whatsoever. Kano also produced a methodology for mapping consumer responses with questionnaires that focused on attractive qualities through reverse qualities. The five categories for customer preferences are as follows:

- Attractive
- One-dimensional
- Must-be
- Indifferent
- Reverse

Attractive qualities are those which provide satisfaction when fulfilled; however do not result in dissatisfaction if not fulfilled.

One-dimensional qualities are those which provide satisfaction when fulfilled, and dissatisfaction if not fulfilled.

Must-be qualities are those which are taken for granted if fulfilled, but provide dissatisfaction when not fulfilled.

Indifferent qualities are those which are neither good nor bad resulting in neither customer satisfaction nor dissatisfaction.

Reverse qualities are those which result in high levels of dissatisfaction from some customers and show that most customers are not alike (Figure 8.41).

A Kano table is shown in Table 8.2.

The Kano model is important to use when being sustainable because it is important to differentiate which aspects we must accomplish to protect our environment and which aspects we can gradually improve upon.

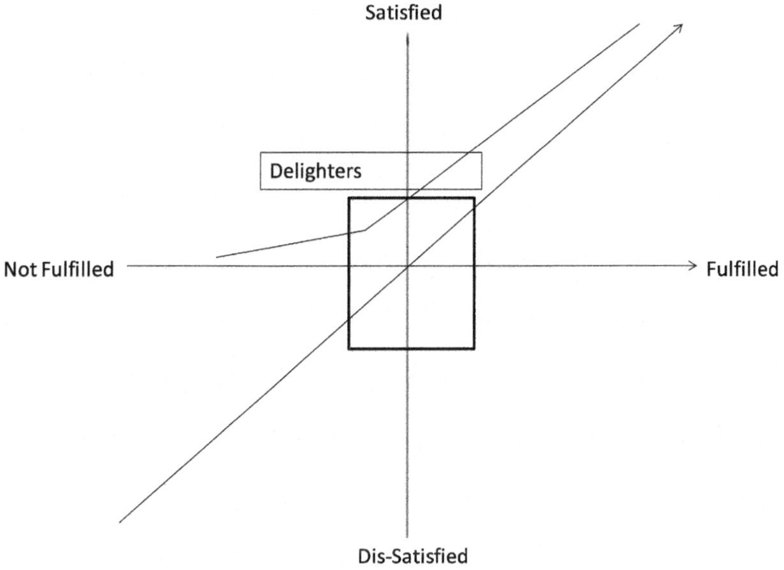

FIGURE 8.41
Kano model.

Critical to Quality

CTQs or Critical to Qualities are characteristics that are important to the customer. They come from the Voice of the Customer (VOC). CTQs are measureable and quantifiable metrics that come from the VOC. An affinity diagram is an organizational tool for VOCs.

Critical to Quality is critical to communication because we need to understand the critical aspects that matter most to the customer.

TABLE 8.2
Kano Model Table

		Answers to a negatively formulated question			
		I like that	That's normal	I don't care	I don't like that
	I like that		Delighter	Delighter	Satisfier
	That's normal				Dissatisfier
	I don't care				Dissatisfier
Answers to a positively formulated question	I don't like that				

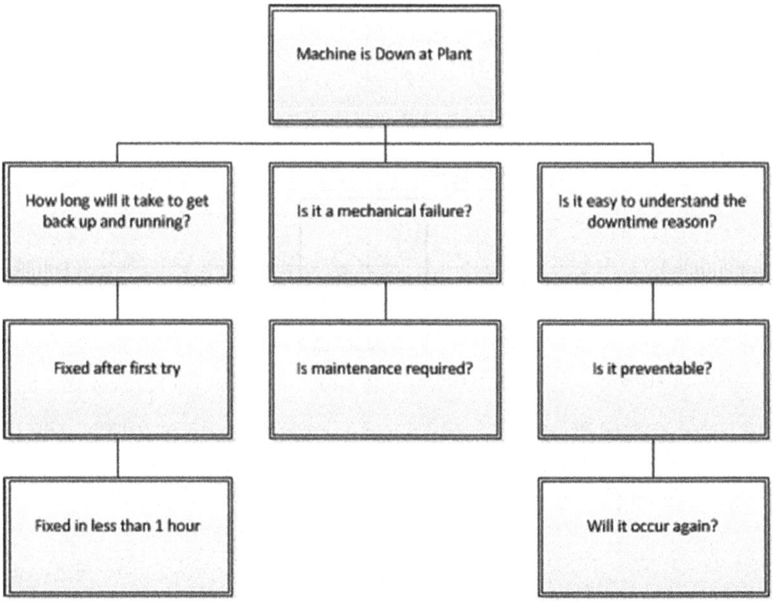

FIGURE 8.42
CTQ tree.

Utilizing a VOC for manufacturing internally is a good way to understand processes that the employees know a great deal about. Therefore, the production worker(s) are the customers and the questions are given to them.

A tree format helps with the visualization of CTQs, as shown in Figure 8.42.

Affinity Diagram

An affinity diagram is a tool conducted to place large amounts of information into an organized manner by grouping the data into characteristics. The steps for an affinity diagram are as follows:

- Step 1 – Clearly define the question or focus of the exercise
- Step 2 – Record all participant responses on note cards or Post-it notes
- Step 3 – Lay out all note cards or post the Post-its onto a wall
- Step 4 – Look for and identify general themes

- Step 5 – Begin moving the note cards or Post-it notes into themes until all responses are allocated
- Step 6 – Re-evaluate and make adjustments

The steps to an affinity diagram for a basic process are shown below:

- Write out notes or information about the process or problem on a piece of paper or sticky note
- Move the notes around so that there are clusters of similar themes in a group (this is normally done by a group)
- Do not explain the thoughts behind the movements of the notes
- Continue the movement of the notes into clusters until everyone seems pleased
- Once this process is completed, the team should discuss the groupings and determine a theme for each grouping
- Problematic reasons can be more visible with the clusters
- Problem-solving can be completed from the clusters

The pros and cons are then sought after to reach a decision. The decision should be made through consensus from the group where the pros outweigh the cons (Figure 8.43).

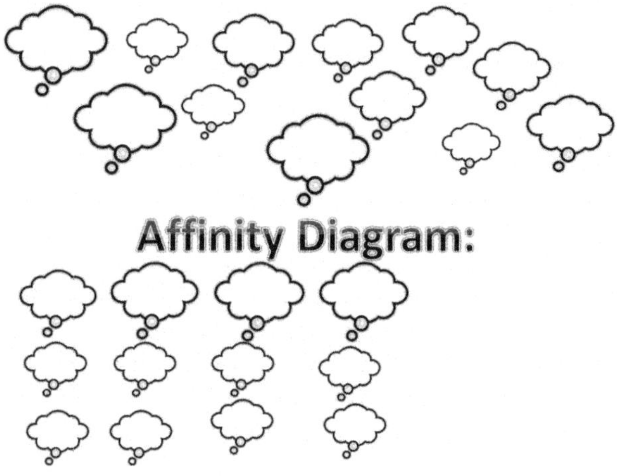

FIGURE 8.43
Affinity diagram example.

Measurement Systems Analysis

Gage R and R

Gage R and R is a measurement systems analysis (MSA) technique that uses continuous data based on the principles that:

- Data must be in statistical control.
- Variability must be small compared to product specifications.
- Discrimination should be about one-tenth of product specifications or process variations.
- Possible sources of process variation are revealed by measurement systems.
- Repeatability and reproducibility are primary contributors to measurement errors.
- The total variation is equal to the real product variation plus the variation due to the measurement system.
- The measurement system variation is equal to the variation due to repeatability plus the variation due to reproducibility.
- Total (observed) variability is an additive of product (actual) variability and measurement variability.

Discrimination is the number of decimal places that can be measured by the system. Increments of measure should be about one-tenth of the width of a product specification or process variation that provides distinct categories.

Accuracy is the average quality near to the true value.

The *true value* is the theoretically correct value.

Bias is the distance between the average value of the measurement and the true value, the amount by which the measurement instrument is consistently off target, or systematic error. *Instrument accuracy* is the difference between the observed average value of measurements and the true value. Bias can be measured based by instruments or operators. Operator bias occurs when different operators calculate different detectable averages for the same measure. Instrument bias results when different instruments calculate different detectable averages for the same measure.

Precision encompasses total variation in a measurement system, the measure of natural variation of repeated measurements, and repeatability and reproducibility.

Repeatability is the inherent variability of a measurement device. It occurs when repeated measurements are made of the same variable under absolutely identical condition (same operators, setups, test units, environmental conditions) in the short term. Repeatability is estimated by the pooled standard deviation of the distribution of repeated measurements and is always less than the total variation of the system.

Reproducibility is the variation that results when measurements are made under different conditions. The different conditions may be operators, setups, test units, or environmental conditions in the long term. Reproducibility is estimated by the standard deviation of the average of measurements from different measurement conditions.

The *measurement capability index* is also known as the precision-to-tolerance (*P*/*T*) ratio. The equation is $P/T = 5.15 \times \sigma$MS)/tolerance. The *P*/*T* ratio is usually expressed as a percent and indicates what percent of the tolerance is taken up by the measurement error. It considers both repeatability and the reproducibility. The ideal ratio is 8% or less; an acceptable ratio is 30% or less. An SD of 5.15 accounts for 99% of MS variation and is an industry standard.

The *P*/*T* ratio is the most common estimate of measurement system precision. It is useful for determining how well a measurement system can perform with respect to the specifications. The specifications, however, may be inaccurate or need adjustment. The %R&R = (σMS/σTotal) × 100 formula addresses the percent of the total variation taken up by measurement error and includes both repeatability and reproducibility.

A Gage R and R can also be performed for discrete data also known as binary data. These data are also known as yes/no or defective/non-defective type data. The data still require at least 30 data points. The percentages of repeatability, reproducibility, and compliance should be measured. If no repeatability is able to be shown, there will also be no reproducibility. The matches should be above 90% for the evaluations. A good measurement system will have a 100% match for repeatability, reproducibility, and compliance.

If the result is below 90%, the operational definition must be re-visited and re-defined. Coaching, teaching, mentoring, and standard operating procedures (SOPs) should be reviewed, and the noise should be eliminated.

A Gage R and R is shown in Figure 8.44 where there is a decision to be made on what equipment is sustainable and what employees are sustainable in a factory based on measurement data.

The Gage R and R bars are desired to be as small as possible, driving the Part-to-Part bars to be larger.

The average of each operator is different, meaning the reproducibility is suspect. The operator is having a problem making consistent measurements.

The Operator*Samples interactions lines should be reasonably parallel to each other. The operators are not consistent to each other.

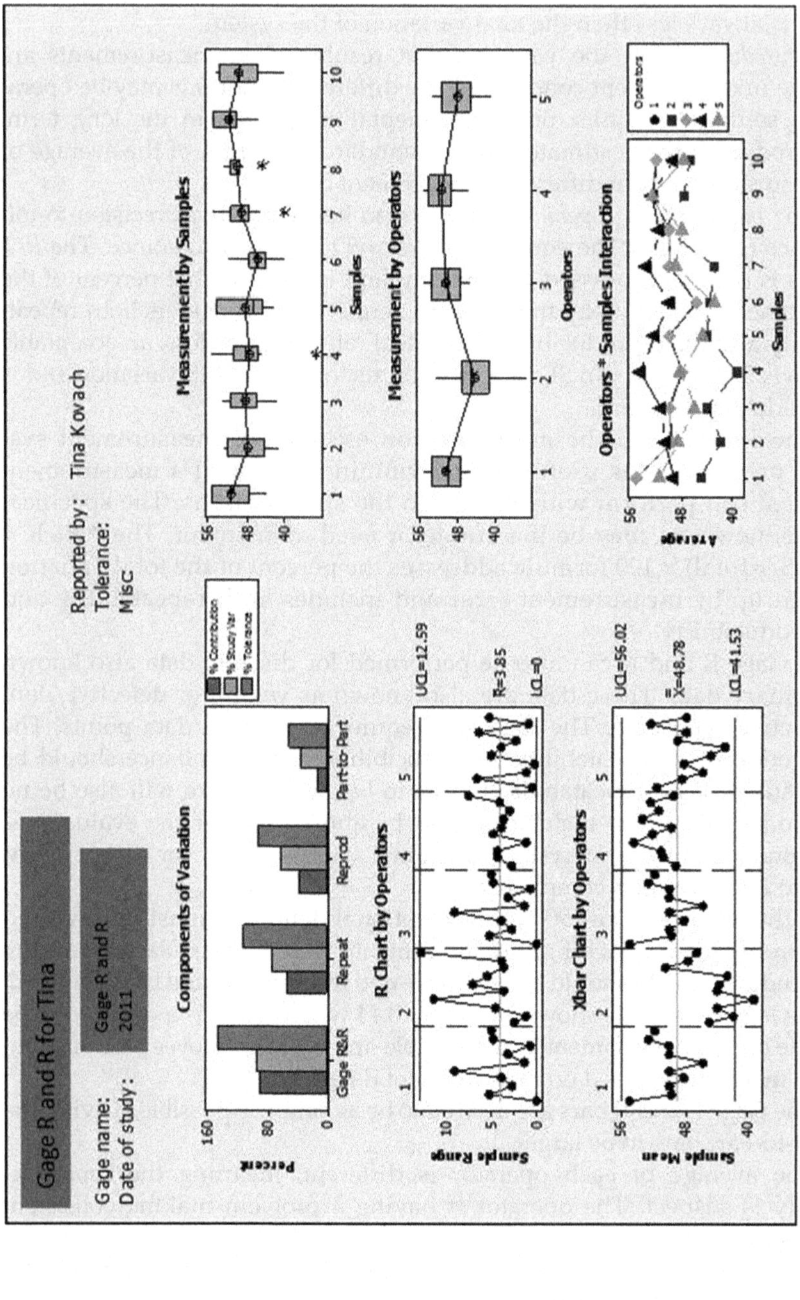

FIGURE 8.44
Gage R and R example.

The Measurement by Samples graph shows there is minimal spread for each sample and a small amount of shifting between samples.

The Measurement by Operators shows that the operators are not consistent, and Operator 2 is normally lower than the rest (Table 8.3).

The Sample by Operator of 0.706 shows that the interaction was not significant which is what is wanted from this study.

TABLE 8.3

Gage R and R Results

Gage R&R Study – ANOVA Method

Gage R&R for Measurement

Gage name: White Butter Creme Gage R and R

Date of study: 11/18/10

Reported by: Tina Kovach

Tolerance:

Misc:

Two-Way ANOVA Table with Interaction

Source	DF	SS	MS	F	P
Samples	9	282.49	31.388	3.3908	0.004
Operators	4	611.14	152.785	16.5050	0.000
Samples * Operators	36	333.25	9.257	0.8398	0.706
Repeatability	50	551.13	11.023		
Total	99	1778.01			

Alpha to remove interaction term = 0.25

Two-Way ANOVA Table without Interaction

Source	DF	SS	MS	F	P
Samples	9	282.49	31.388	3.0523	0.003
Operators	4	611.14	152.785	14.8573	0.000
Repeatability	86	884.38	10.283		
Total	99	1778.01			

Gage R&R

%Contribution		
Source	VarComp	(of VarComp)
Total Gage R&R	17.4086	89.19
Repeatability	10.2835	52.68
Reproducibility	7.1251	36.50
Operators	7.1251	36.50
Part-To-Part	2.1104	10.81
Total Variation	19.5190	100.00

Process tolerance = 15

Study Var	%Study Var	%Tolerance		
Source	StdDev (SD)	(5.15 * SD)	(%SV)	(SV/Toler)
Total Gage R&R	4.17236	21.4876	94.44	143.25
Repeatability	3.20679	16.5150	72.58	110.10
Reproducibility	2.66929	13.7468	60.42	91.65
Operators	2.66929	13.7468	60.42	91.65
Part-To-Part	1.45274	7.4816	32.88	49.88
Total Variation	4.41803	22.7529	100.00	151.69

Number of Distinct Categories = 1

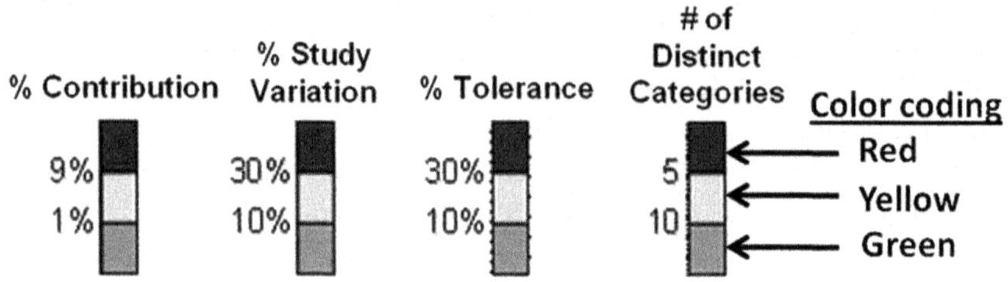

FIGURE 8.45
Gage R and R results.

In a Gage R&R, the % contribution for the part to part of 10.81 shows the parts are the same.

The total Gage R and R% study variation of 94.44, % contribution of 89.19, tolerance of 143.25, and distinct categories of 1 showed that there was no repeatability, reproducibility, and was not a good Gage. The number of categories being less than two shows the measurement system is of minimal value since it will be difficult to distinguish one part from another.

The Gage is a bad representation based on Figures 8.45 and 8.46.

The Gage Run Chart shows that there is no consistency between measurements.

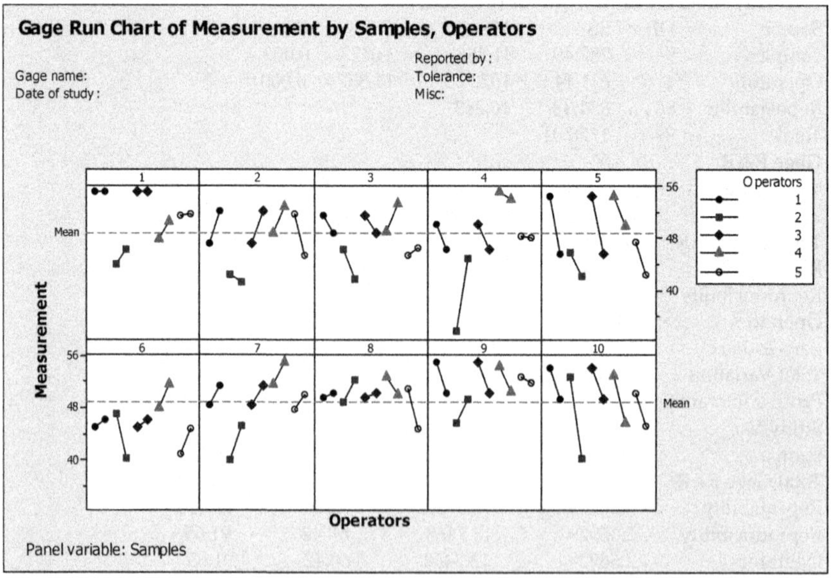

FIGURE 8.46
Gage R and R run chart.

There is no reproducibility or repeatability between any of the measurements.

Process Capabilities

The capability of a process is the spread that contains most of the values of the process distribution. Capability can only be established on a process that is stable with a distribution that only has common cause variation (Figure 8.47).

Process Capability Example

Capable process (C_p)

A process is capable ($C_p \geq 1$) if its natural tolerance lies within the engineering tolerance or specifications. The measure of process capability of a stable process is $6\hat{\sigma}$, where $\hat{\sigma}$ is the inherent process variability that is estimated from the process. A minimum value of $C_p = 1.33$ is generally used for an ongoing process. This ensures a very low reject rate of 0.007%, and therefore is an effective strategy for prevention of nonconforming items. C_p is defined mathematically as

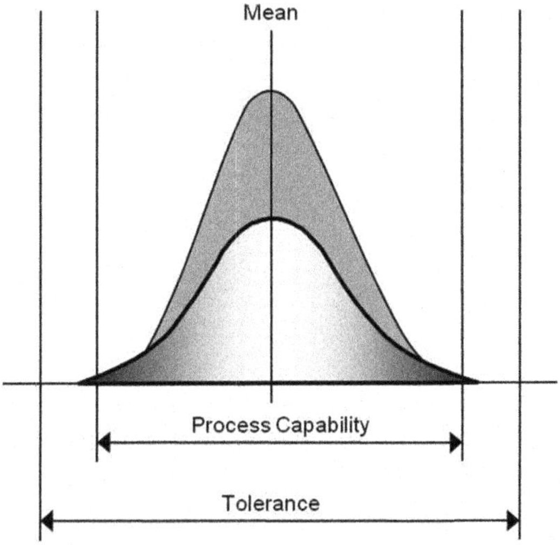

FIGURE 8.47
Process capability.

$$C_p = \frac{USL - LSL}{6\hat{\sigma}}$$

$$= \frac{\text{allowable process spread}}{\text{actual process spread}}$$

where
USL = upper specification limit
LSL = lower specification limit

C_p measures the effect of the inherent variability only. The analyst should use R-bar/d_2 to estimate $\hat{\sigma}$ from an R-chart that is in a state of statistical control, where R-bar is the average of the subgroup ranges and d_2 is a normalizing factor that is tabulated for different subgroup sizes (n). We don't have to verify control before performing a capability study. We can perform the study, then verify control after the study with the use of control charts. If the process is in control during the study, then our estimates of capabilities are correct and valid. However, if the process was not in control, we would have gained useful information, as well as proper insights as to the corrective actions to pursue.

Capability Index (C_{pk})

Process centering can be assessed when a two-sided specification is available. If the capability index (C_{pk}) is equal to or greater than 1.33, then the process may be adequately centered. C_{pk} can also be employed when there is only one-sided specification. For a two-sided specification, it can be mathematically defined as:

$$C_{pk} = \text{Minimum}\left\{\frac{USL - \bar{X}}{3\hat{\sigma}}, \frac{\bar{X} - LSL}{3\hat{\sigma}}\right\}$$

where
$\bar{X}$ = Overall process average

However, for a one-sided specification, the actual C_{pk} obtained is reported. This can be used to determine the percentage of observations out of specification. The overall long-term objective is to make C_p and C_{pk} as large as possible by continuously improving or reducing process variability, $\hat{\sigma}$, for every iteration so that a greater percentage of the product is near the key quality characteristics target value. The ideal is to center the process with zero variability.

If a process is centered but not capable, one or several courses of action may be necessary. One of the actions may be that of integrating designed experiment to gain additional knowledge on the process and designing control strategies. If excessive variability is demonstrated, one may conduct a nested design with the objective of estimating the various sources of variability. These sources of variability can then be evaluated to determine what strategies to use in order to reduce or permanently eliminate them. Another action may be that of changing the specifications or continuing production and then sorting the items. Three characteristics of a process can be observed with respect to capability, as summarized below:

1. The process may be centered and capable
2. The process may be capable but not centered
3. The process may be centered but not capable

Possible Applications of Process Capability Index

The potential applications of process capability index are summarized below:

- *Communication* – C_p and C_{pk} have been used in industry to establish a dimensionless common language useful for assessing the performance of production processes. Engineering, quality, manufacturing, etc., can communicate and understand processes based on the process capabilities.
- *Continuous improvement* – The indices can be used to monitor continuous improvement by observing the changes in the distribution of process capabilities. For example, if there were 20% of processes with capabilities between 1 and 1.67 in a month, and some of these improved to between 1.33 and 2.0 the next month, then this is an indication that improvement has occurred.
- *Audits* – There are so many various kinds of audits in use today to assess the performance of quality systems. A comparison of in-process capabilities with capabilities determined from audits can help establish problem areas.
- *Prioritization of improvement* – A complete printout of all processes with unacceptable C_p or C_{pk} values can be extremely powerful in establishing the priority for process improvements.
- *Prevention of nonconforming product* – For process qualification, it is reasonable to establish a benchmark capability of $C_{pk} = 1.33$ which will make nonconforming products unlikely in most cases.

Potential Abuse of C_p and C_{pk}

In spite of its several possible applications, process capability indices have some potential sources of abuse as summarized below:

- *Problems and drawbacks* – C_{pk} can increase without process improvement even though repeated testing reduces test variability. The wider the specifications, the larger the C_p or C_{pk}, but the action does not improve the process.
- Analysts tend to focus on number rather than on process.
- *Process control* – Analysts tend to determine process capability before statistical control has been established. Most people are not aware that capability determination is based on process common cause variation and what can be expected in the future. The presence of special causes of variation makes prediction impossible and capability index unclear.
- *Non-normality* – Some processes result in non-normal distribution for some characteristics. Since capability indices are very sensitive to departures from normality, data transformation may be used to achieve approximate normality.
- *Computation* – Most computer-based tools do not use $\overline{R} / d_2$ to calculate σ.

When analytical and statistical tools are coupled with sound managerial approaches, an organization can benefit from a robust implementation of improvement strategies. One approach that has emerged as a sound managerial principle is Lean, which has been successfully applied to many industrial operations.

C_p and C_{pk} are capability analyses that can only be done with normal data. It is very easy to use any data for capability analyses especially on software systems that will calculate the data automatically. The first step in doing the capability analysis is to have continuous data and check for normality. Only if the data are normal, should capability studies be performed. If the data are not normal, the special cause variation is sought after. Data points may only be taken out if the reasoning is known for the data point that is an outlier (e.g. temperature change, shift change). Once an outlier is found for a known reason, the outlier can be removed and the data can be checked for normality once again. If there is no root cause for the outlier, more data must be taken, but capability analyses should not be done until the normality is proven.

The importance of finding the capable equipment or products in a business through process capabilities will allow the variation to be found through benchmarking. The best processes should be used for these benchmarking techniques. The best–in-class (BIC) practices should be performed on the different equipments/products/processes. Then improvements should be made on the areas that are not as capable. It is very important to perform preventative maintenance on all and any equipment in order for the equipment to stay performing at the highest possible process capability.

When two or more equipment pieces are being compared, the first step is to perform a normality test, as stated in Figures 8.48–8.51.

The conclusions that come from the normality tests are the following:

Machine A1 is not normal

Machine B1 is the most normal

Machine B2 is not normal

Small mixer is JUST in at normal

The normal pieces of equipment are then checked for the process capabilities (Figures 8.52 and 8.53).

Machine B1 is your best machine, the short-term capability (C_p of 1.34) is approximately equivalent to a short-term Z of 4, which is good. The long-term capability needs some improvement (P_{pk} of 0.94).

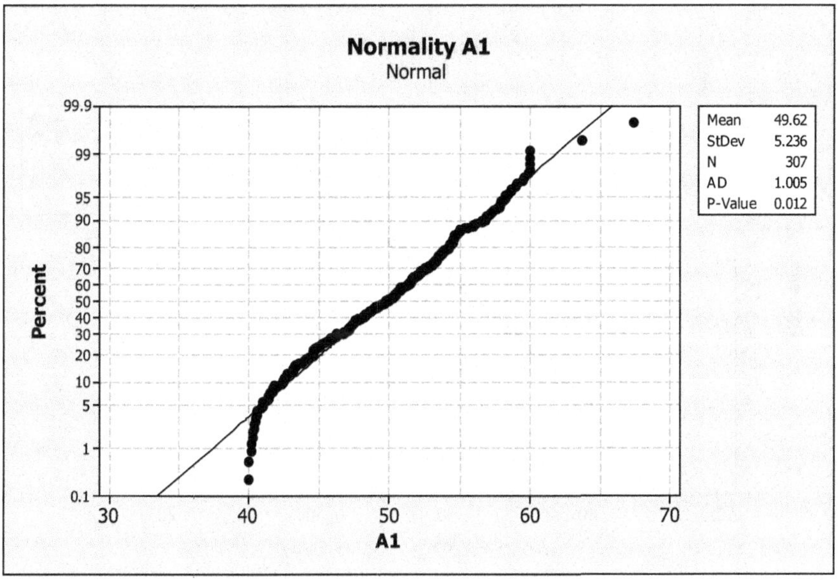

FIGURE 8.48
Normality of machine A1.

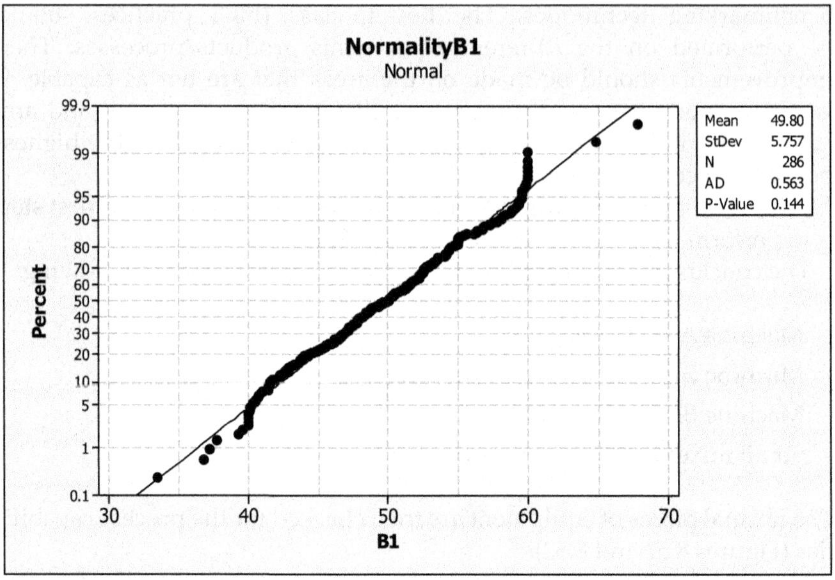

FIGURE 8.49
Normality of machine B1.

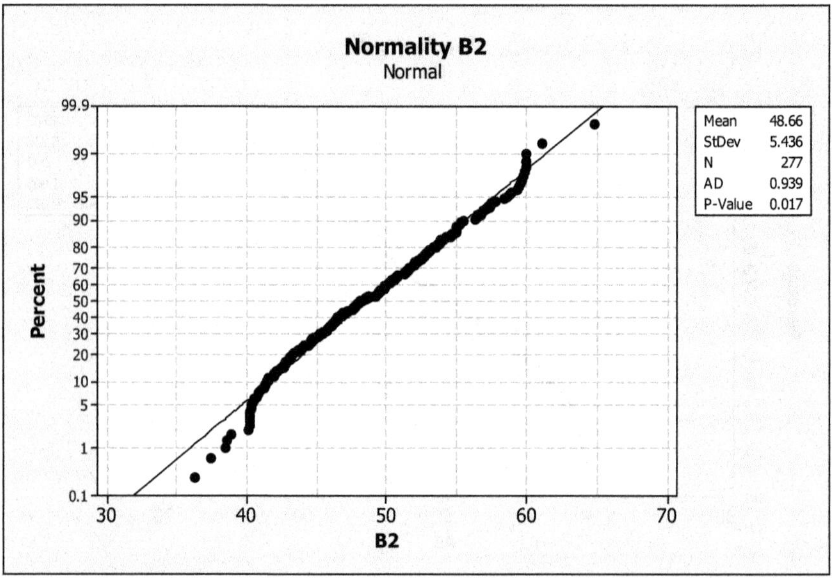

FIGURE 8.50
Normality of machine B2.

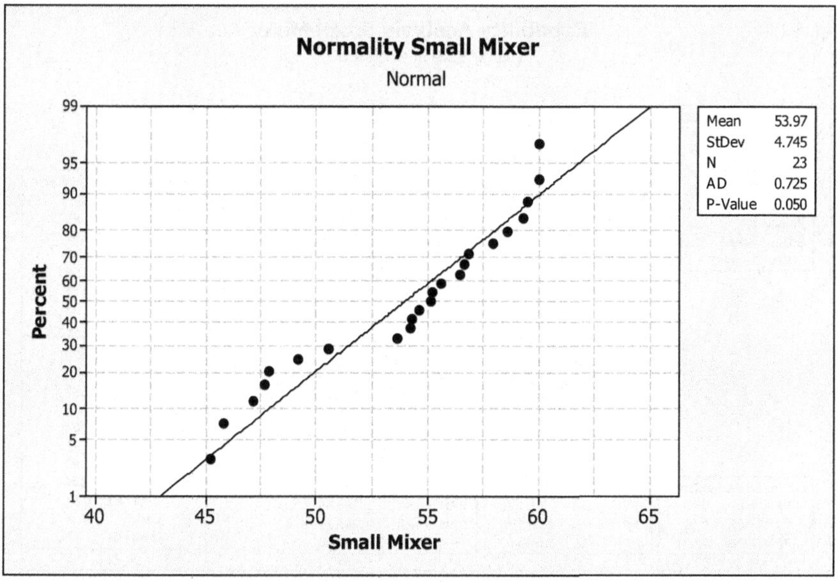

FIGURE 8.51
Normality of equipment small mixer.

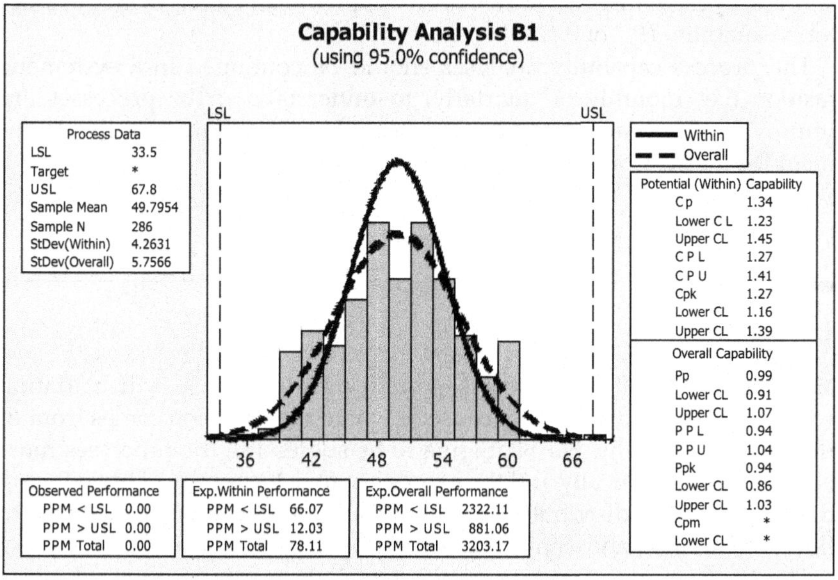

FIGURE 8.52
Process capability B1.

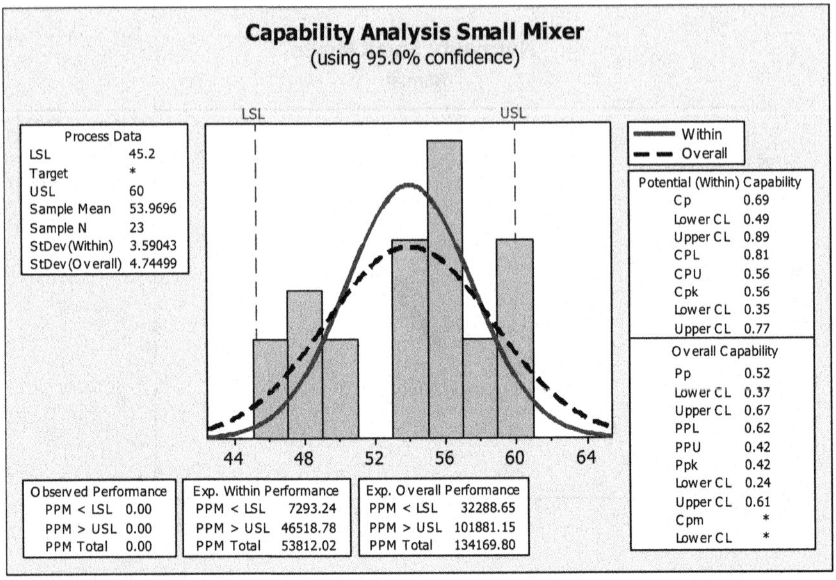

FIGURE 8.53
Process capability small mixer.

Small mixer is just normal, but is better than Machine A1 or B2. The short-term capability (C_p of 0.69) needs improvement along with the long-term capability (P_{pk} of 0.42).

The process capability analyses should be continued in a systematic fashion (i.e., monthly or quarterly) to understand if the processes are improving. Continuous improvement should be performed on the equipment for the best capabilities.

Variation

Variation is present in all processes, but the goal is to reduce the variation while understanding the root cause of where the variation comes from in the process and why. For Six Sigma to be successful, the processes must be in control statistically and the processes must be improved by reducing the variation. The distribution of the measurements should be analyzed to find the variation and depict the outliers or patterns.

The study of variation began with Dr. W. Edwards Deming who was also known as the Father of Statistics. Deming stated that variation happens naturally, but the purpose is to utilize statistics to show patterns and types of variations. There are two types of variations that are sought after: special cause variation and common cause variation. Special cause variation refers to out of the ordinary events such as a power outage, whereas common cause variation is inherent in all processes and is typical. The variation is sought to be reduced so that the processes are predictable, in statistical control, and have a known process capability. A root cause analysis should be done on special cause variation so that the occurrence is not to happen again. Management is in charge of common cause variation where action plans are given to reduce the variation.

Assessing the location and spread are important factors as well. Location is known as the process being centered along with the process requirements. Spread is known as the observed values compared to the specifications. The stability of the process is required. The process is said to be in statistical control if the distribution of the measurements has the same shape, location, and spread over time. This is the point in time where all special causes of variation are removed and only common cause variation is present.

An *average, central tendency* of a data set is a measure of the "middle" or "expected" value of the data set. Many different descriptive statistics can be chosen as measurements of the central tendency of the data items. These include the arithmetic mean, the median, and the mode. Other statistical measures such as the standard deviation and the range are called measures of spread of data. An average is a single value meant to represent a list of values. The most common measure is the arithmetic mean, but there are many other measures of central tendency such as the median (used most often when the distribution of the values is skewed by small numbers with very high values).

As stated before, special cause variation would be occurrences such as power outages and large mechanical breakdowns. Common cause variations would be occurrences such as electricity being different by a few thousand kilowatts per month. In order to understand the variation, graphical analyses should be done followed by capability analyses.

It is important to understand the variation in the systems so that the best performing equipment is used. The variation sought after is in turn utilized for sustainability studies. The best performing equipment should be utilized the most and the least performing equipment should be brought back to its original state of condition and then upgraded or fixed to be capable. Graphical analysis is explained next.

Graphical Analysis

Graphical analyses are visual representations of tools that show meaningful key aspects of projects. These tools are commonly known as dot

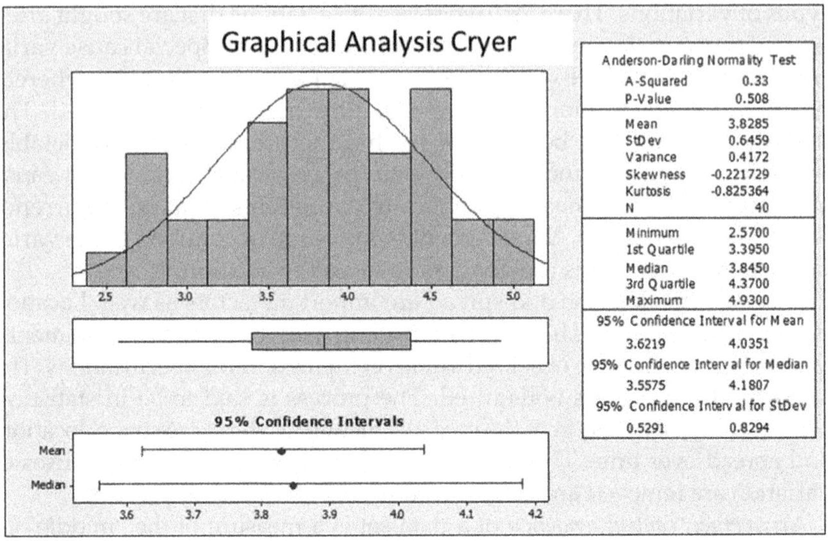

FIGURE 8.54
Graphical analysis.

plots, histograms, normality plots, Pareto diagrams, second-level Pareto's (also known as stratification), box plots, scatter plots, and marginal plots. The plotting of data is a key beginning step to any type of data analysis because it is a visual representation of the data.

A graphical analysis can be shown in Figure 8.54.

A graphical analysis summary can cover sample size, mean, standard deviation, variance, skewness, kurtosis, *p*-value, and confidence intervals as shown in Figure 8.54.

A Pareto chart shows a visual representation of what occurs the most as shown in Figure 8.55.

Cause and Effect Diagram

After a process is mapped, the cause and effect (C&E) diagram can be completed. This process is so important because it enables root cause analysis. The basis behind root cause analysis is to ask, "Why?" five times in order to get to the actual root cause. Many times problems are "band-aided" in order to fix the top-level problem, but the actual problem itself is not addressed.

A cause and effect diagram is shown in Figure 8.56.

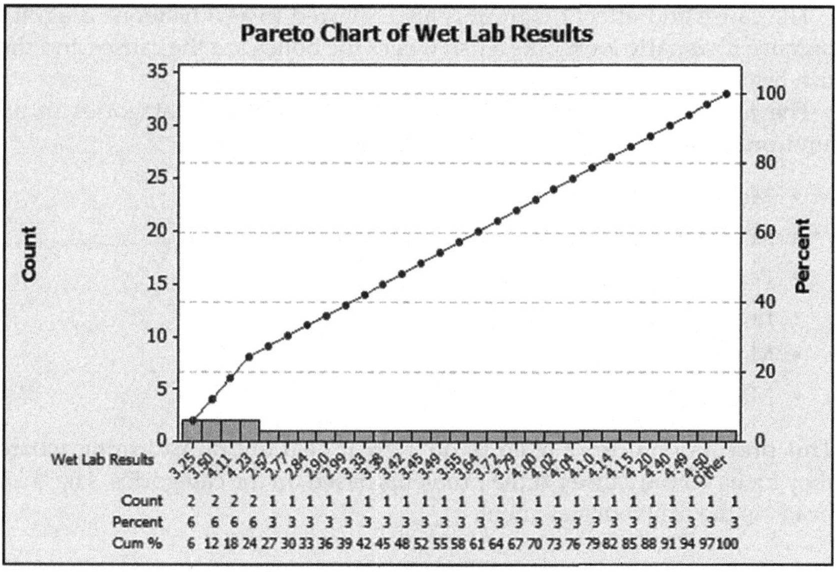

FIGURE 8.55
Pareto chart.

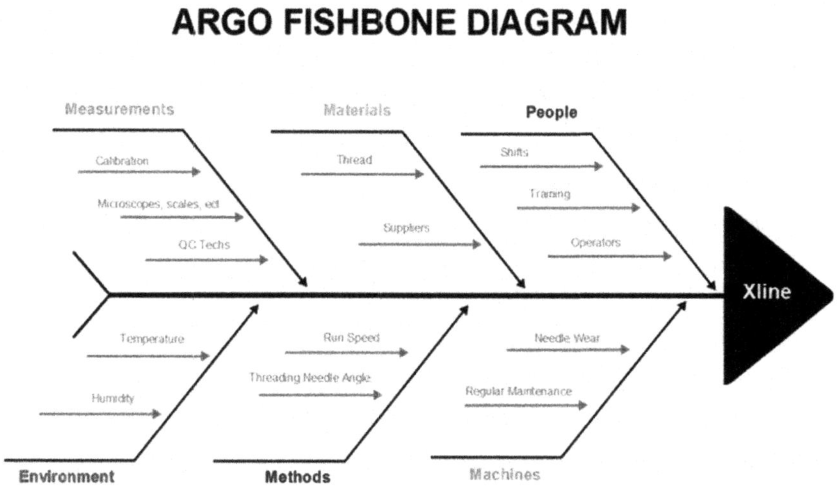

FIGURE 8.56
Cause and effect diagram.

The cause and effect diagram is also referred to as a fishbone diagram because it visually looks like a fish where the bones are the causes and the fish head is the effect.

The fish bone is broken out to the most important categories in an environment:

- Measurements
- Material
- Personnel
- Environment
- Methods
- Machines

This process requires a team to do a great deal of brainstorming where they focus on the causes of the problems based on the categories. The "fish head" is the problem statement.

Failure Mode and Effect Analysis

In order to select action items from the C & E diagram and prioritize the projects, failure mode and effect analyses (FMEAs) are completed. The FMEA will identify the causes, assess risks, and determine further steps. The steps to an FMEA are the following:

1. Define process steps
2. Define functions
3. Define potential failure modes
4. Define potential effects of failure
5. Define the severity of a failure
6. Define the potential mechanisms of failure
7. Define current process controls
8. Define the occurrence of failure
9. Define current process control detection mechanisms
10. Define the ease of detecting a failure

TABLE 8.4

Detection Criteria

Detection	Criteria: Likelihood the Existence of a Defect Will Be Detected by Test Content before Product Advances to Next or Subsequent Process	Ranking
Almost Impossible	Test content detects < 80% of failures	10
Very Remote	Test content must detect 80% of failures	9
Remote	Test content must detect 82.5% of failures	8
Very Low	Test content must detect 85% of failures	7
Low	Test content must detect 87.5% of failures	6
Moderate	Test content must detect 90% of failures	5
Moderately High	Test content must detect 92.5% of failures	4
High	Test content must detect 95% of failures	3
Very High	Test content must detect 97.5% of failures	2
Almost Certain	Test content must detect 99.5% of failures	1

11. Multiply severity, occurrence, and detection to calculate a risk priority number (RPN)
12. Define recommended actions
13. Assign actions with key target dates to responsible personnel
14. Revisit the process after actions have been taken to improve it
15. Recalculate RPNs with the improvements

The detection, severity, and occurrence criteria are shown in Tables 8.4–8.6.

An FMEA is shown in Figure 8.57.

What can be seen from the FMEA that is an important aspect to sustainability is the RPN number reducing after the action items. It is important to understand the processes severity to a customer and increasing the capability of the process in turn improves the process. The RPNs reducing will make the entire process more sustainable by being able to deliver the process at the best capabilities through thorough project management. It is important to maintain the FMEA so that once a process is improved, it is documented.

TABLE 8.5

Severity Criteria

Effect	Criteria: Severity of the Effect	Ranking
Hazardous – Without Warning	Very high severity ranking when a potential failure mode affects safety and involves noncompliance without warning	10
Hazardous – With Warning	Very high severity ranking when a potential failure mode affects safety and involves noncompliance with warning	9
Very High	Process is not operable and has loss of its primary function	8
High	Process is operable, but with a reduced functionality and an unhappy customer	7
Moderate	Process is operable, but not easy to manufacture. The customer is uncomfortable	6
Low	Process is operable, but uncomfortable with a reduced level of performance. The customer is dissatisfied	5
Very Low	The process is not in 100% compliance. Most customers are able to notice the defect	4
Minor	The process is not in 100% compliance. Some customers are able to notice the defect	3
Very Minor	The process is not in 100% compliance. Very few customers are able to notice the defect	2
None	No effect	1

TABLE 8.6

Occurrence Criteria

Probability of Failure	Possible Failure Rates	Ranking
Failures is almost inevitable	≥1 in 2	10
	1 in 3	9
High: Repeated Failures	1 in 8	8
	1 in 20	7
Moderate: Occasional Failures	1 in 80	6
	1 in 400	5
	1 in 2000	4
Low: Very Few Failures	1 in 15,000	3
	1 in 150,000	2
Remote: Failure is Unlikely	≤1 in 1,500,000	1

Process Function (Step)	Potential Failure Modes (process defects)	Potential Failure Effects (KPOVs)	SEV	Class	Potential Causes of Failure (KPIVs)	OCC	Current Process Controls	DET	RPN	Recommend Actions	Responsible Person & Target Date	Taken Actions	SEV	OCC	DET	RPN
Copper strikes	Agitator	100% Down, potential over reactor	9		Motor, Bearing, Shaft, Gearbox	3	None	5	135							
APV	Product Collector Roto Lock	Product not discharging	8		Motor bad, communication with scale, jams	4	None	2	64							
APV	Blowers	100% Down	8		Plugged filter, bad motor, bad coupling	3	PM on blowers every 4 months	2	48							
Copper nitrate makeup	Discharge valve	Pluggage	8		Powder build up before dissolved	2	Valve design - flush mount	3	48							
Copper nitrate makeup	Mag drive pump	Won't pump	8		Running dry, worn out, motor failure	6	Level indication, re-circulation	1	48	Load monitor to be put on, re-dundant pump						
Copper nitrate makeup	Gate failure	100% Down	8		Damage	6	None	1	48	Limit switch, investigate new gate						

FIGURE 8.57
Failure modes and effects matrix.

Hypothesis Testing

Hypothesis testing validates assumptions made by verification of the processes based on statistical measures. It is important to use at least 30 data points for hypothesis testing so that there is enough data to validate the results.

Normality of the data points must be found in order for the hypothesis testing to be accurate.

The assumptions are shown in the null and alternate hypothesis:

Ho = (The null hypothesis): The difference is equal to the chosen reference value $\mu_1-\mu_2=0$.

Ha = (The alternate hypothesis): The difference is not equal to the chosen reference value $\mu_1-\mu_2$ is not =0.

An example of a paired *t*-test for hypothesis testing is shown in Table 8.7.

95% confidence interval (CI) for mean difference: (1.16, 6.69) *t*-test of mean difference = 0 (vs not = 0): *t*-Value = 2.90 *p*-value = 0.007.

TABLE 8.7

Hypothesis Testing Paired *t*-Test Example

Paired T for Before – After	N	Mean	StDev	SE Mean
Before	30	83.623	5.195	0.948
After	30	79.697	4.998	0.913
Difference	30	3.93	7.41	1.35

The confidence interval for the mean difference between the two materials does not include zero, which suggests a difference between them. The small *p*-value ($p = 0.007$) further suggests that the data are inconsistent with H_0: $\mu\ d = 0$, that is, the two materials do not perform equally. Specifically, the first set (mean = 79.697) performed better than the next set (mean = 83.623) in terms of weight control over the time span. Conclusion, Reject H_0, the difference is not equal to the chosen reference value: $\mu_1-\mu_2$ is not = 0. The histogram of differences and the boxplot of differences are graphed in Figures 8.58 and 8.59, respectively.

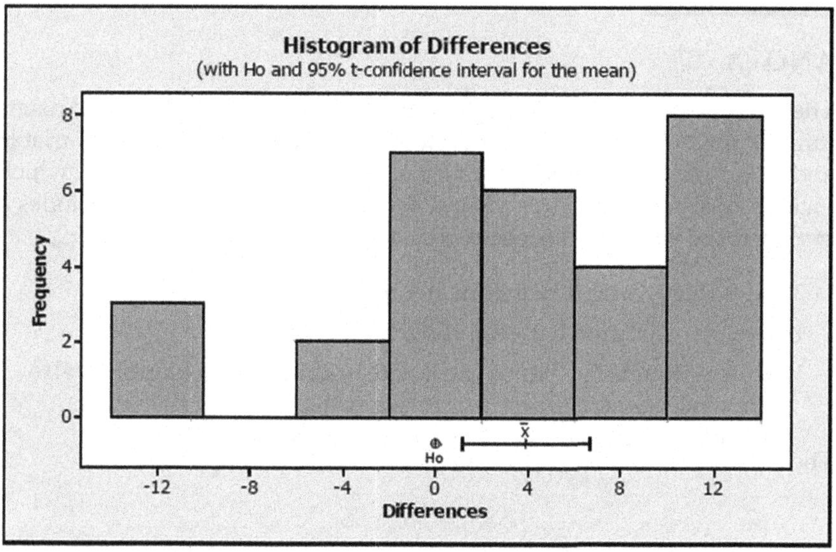

FIGURE 8.58
Histogram of differences.

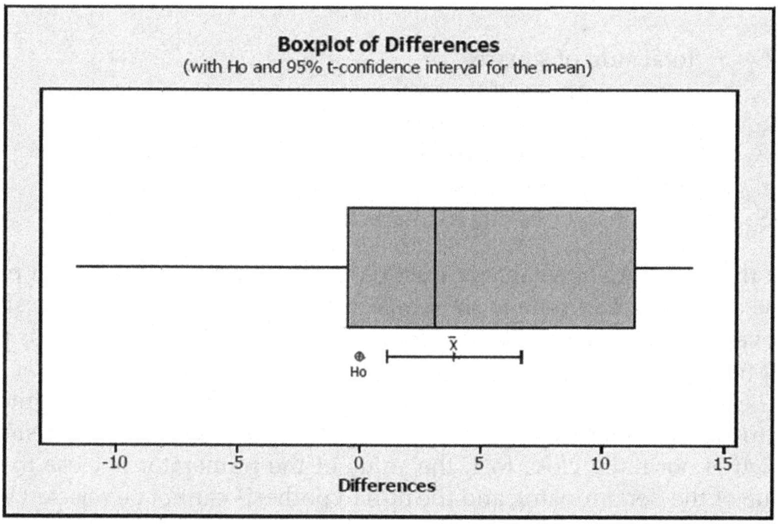

FIGURE 8.59
Boxplot of differences.

ANOVA

The purpose of an ANOVA, also known as analysis of variance, is to determine if there is a relationship between a discrete independent variable and a continuous dependent output. There is a one-way ANOVA which includes one-factorial variance and a two-way ANOVA which includes a two-factorial variance. Three sources of variability are sought after:

Total – Total variability within all observations

Between – Variation between subgroup means

Within – Random chance variation within each subgroup, also known as noise

The equation for a one-way ANOVA is:

$$SS_T = SS_F + SS_e$$

The principles for the one-way ANOVA and two-way ANOVA are the same except that in a two-way ANOVA, the factors can take on many levels. The total variability equation for a two-way ANOVA is:

$$SS_T = SS_A + SS_B + SS_{AB} + SS_e$$

where

SS_T = total sum of squares
SS_F = sum of squares of the factor
SS_e = sum of squares from error
SS_A = sum of squares for factor A
SS_B = sum of squares for factor B
SS_{AB} = sum of squares due to interaction of factors A and B

If the ANOVA shows that at least one of the means is different, a pairwise comparison is done to show which means are different. The residuals, variance, and normality should be examined and the main effects plot and interaction plots should be generated.

The *F*-ratio in an ANOVA compares the denominator to the numerator to see the amount of variation is expected. When the F-ratio is small, which is normally close to 1, the value of the numerator is close to the value of the denominator, and the null hypothesis cannot be rejected stating the numerator and denominator are the same. A large F-ratio indicates the numerator and denominator are different, also known as the MS error, where the null hypothesis is rejected.

Outliers should also be sought after in the ANOVA showing the variability is affected.

The main effects plot shows the mean values for the individual factors being compared. The differences between the factor levels can be seen with the slopes in the lines. The *p*-values can help determine if the differences are significant.

Interaction plots show the mean for different combinations of factors.

Correlation

The linear relationship between two continuous variables can be measured through correlation coefficients. The correlation coefficients are values between −1 and 1.

If the value is around 0, there is no linear relationship.

If the value is less than .50, there is a weak correlation.

If the value is less than .80, there is a moderate correlation.

If the value is greater than .80, there is a strong correlation.

If the value is around 1, there is a perfect correlation.

Simple Linear Regression

Regression analysis describes the relationship between a dependent and an independent variable as a function $y = f(x)$.

The equation for simple linear regression as a model is:

$$Y = b_0 + b_1 x + {}_E$$

Y is the dependent variable
b_0 is the axis intercept
b_1 is the gradient of the regression line
x is the independent variable
E is the error term or residuals

The predicted regression function is tested with the following formula: Signal-Noise Ratio

$$R^2 = \mathrm{SS}_{\mathrm{Model}} / \mathrm{SS}_{\mathrm{Total}}$$

Coefficient of determination, describes the percent of the variation in Y that can be predicted by knowing the change in X.

$$SS_{total} = SS_{model} + SS_{error}$$

Note: When the no constant option is selected, the total sum of square is uncorrected for the mean. Thus, the R^2 value is of little use, since the sum of the residuals is not zero.

The *F*-test shows if the predicted model is valid for the population and not just the sample. The model is statistically significant if the predicted model is valid for the population.

The regression coefficients are tested for significance through *t*-tests with the following hypothesis:

H_0: $b_0 = 0$, the line intersects the origin

H_A: $b_0 \neq 0$, the line does not intersect the origin

H_0: $b_1 = 0$, there is no relationship between the independent variable x_i and the dependent variable y

H_A: $b_1 \neq 0$, there is a relationship between the independent variable x_i and the dependent variable y

A fitted line plot can be done to see the inverse relationship as shown in Figure 8.60.

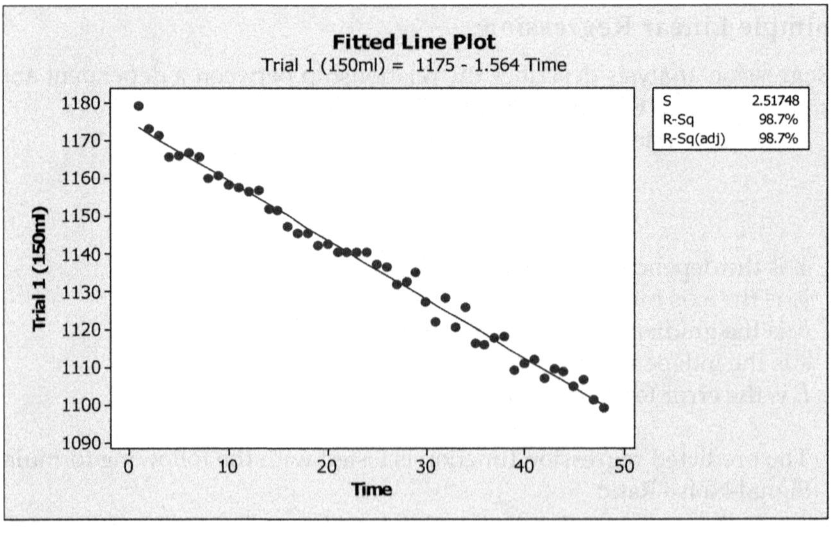

FIGURE 8.60
Fitted line plot.

After the inverse relationship is seen, a regression analysis can be performed.

An example is shown below for the analysis of whether there was a pressure degradation over time on a particular piece of equipment.

A linear relationship was sought after. First, it was sought to see if there was correlation since it can be seen that there is a linear relationship between the variables. The y variable was the measurement and the x variable was the time.

Correlations: Trial 1, Time

Pearson correlation of Trial 1 (150 ml) and Time = −0.994

p-value = 0.000

This correlation coefficient of $r = -0.994$ shows a high, positive dependence.

The p-value being less than 0.05 also shows that the correlation coefficient is significant.

The regression equation is:

$$\text{Trial } 1 = 1175 - 1.56 \text{ Time}$$

Predictor	Coef	SE Coef	T	P
Constant	1174.89	0.74	1591.47	0.000
Time	–1.56360	0.02623	–59.61	0.000

S = 2.51748 R-Sq = 98.7% R-Sq(adj) = 98.7%

Analysis of Variance

Source	DF	SS	MS	F	P
Regression	1	22522	22522	3553.63	0.000
Residual Error	46	292	6		
Total	47	22813			

Unusual Observations

		Trial 1				
Obs	Time	(150ml)	Fit	SE Fit	Residual	St Resid
1	1.0	1179.00	1173.33	0.72	5.67	2.35R
29	29.0	1135.10	1129.55	0.38	5.55	2.23R

R denotes an observation with a large standardized residual.

For each time, there is 1.5 measurement of degradation according to the equation:

$$Y_1 = \beta_0 + \beta_1 X_1$$

The slope equals 1.564. There is a negative slope.

In this particular situation, it becomes critical after losing more than 10% of the measurement in the specification range. According to the graph, about every ten trials, there is a degradation of about 15.

Hypothesis Testing

The *p*-value shows that this is not normal.

H_0 = Accept null hypothesis ($\beta = 0$, no correlation)
H_a = Reject null hypothesis ($\beta \neq 0$, there is correlation)

Therefore, reject null hypothesis. There is correlation.

In the above, the R^2 value is 98.7% which means:

- 98.7% of the *Y* variable's (pressure) can be explained by the model (the regression equation)

The residuals are then evaluated shown in Figure 8.61.

The normality is also taken of the residuals shown in Figure 8.62.

The normality test passes with a value of 0.850. The residuals are in control (Figure 8.63).

The residuals are contained in a straight band, with no obvious pattern in the graph showing that this model is adequate.

Conclusion: Reject H_0, the slope of the line does not equal 0. There is a linear relationship in the measurement versus time, showing that there is correlation. This model proves to be adequate due to the testing done above.

Theory of Constraints

Dr. Eliyahu M. Goldratt created a theory of constraints (TOC). This management theory proved that every systems has at least one constraint limiting it from 100% efficiency. The analysis of a system will show the

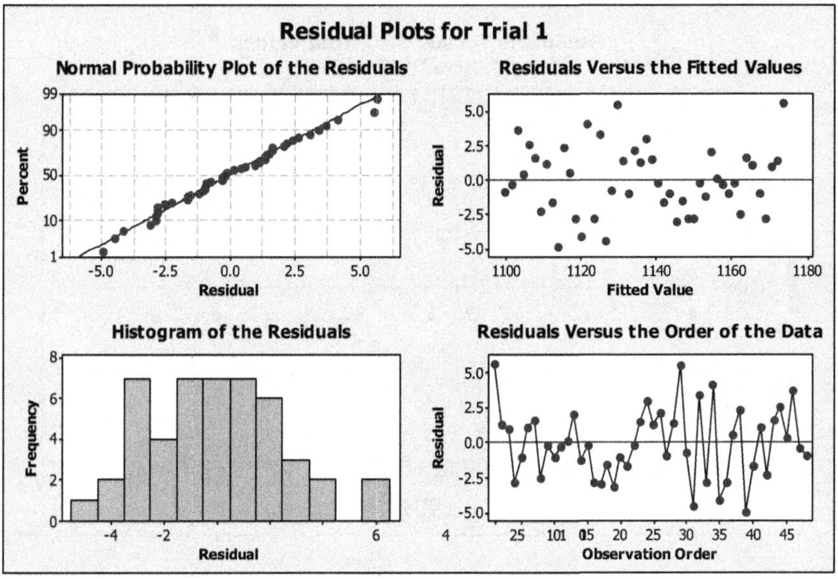

FIGURE 8.61
Residual plots.

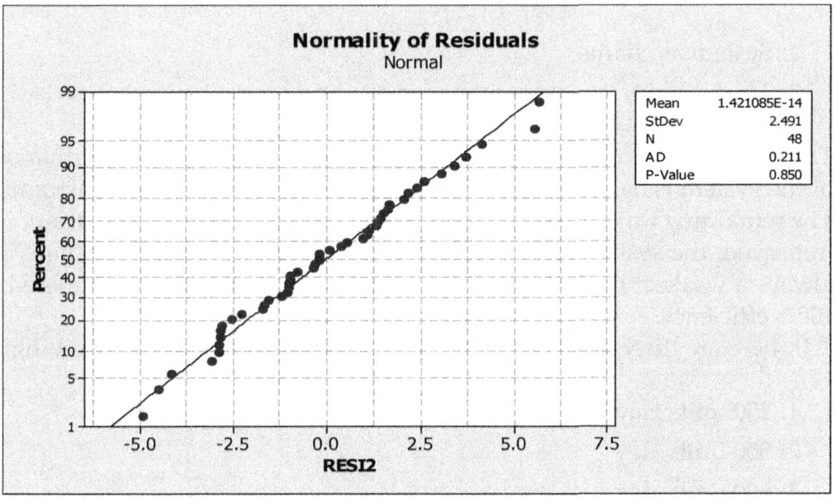

FIGURE 8.62
Normality of residuals.

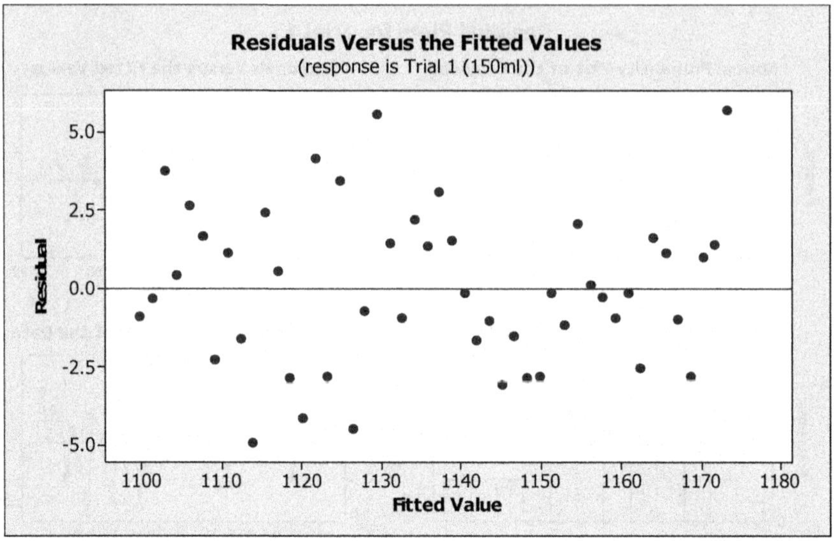

FIGURE 8.63
Residuals versus fitted values.

boundaries of the system. TOC not only shows the cause of the constraints, but it also provides a way to resolve the constraints. There are two underlying concepts with TOC:

1. System as chains
2. Throughput, inventory management, and operating expenses

The performance of the entire system is called the chain. The performance of the system is based on the weakest link of the chain or the constraint. The remaining links are known as non-constraints. Once the constraint is improved, the system becomes more productive or efficient, but there is always a weakest link or constraint. This process continues until there is 100% efficiency.

If there are three manufacturing lines and they produce the following:

1. 250 units/day
2. 500 units/day
3. 600 units/day

The weakest link is manufacturing line 1 because it produces the least amount of units/day. The weakest link is investigated until it reaches the capacity of the non-constraints. After the improvement has been made, the new weakest link is investigated until the full potential of the manufacturing lines can be fulfilled without exceeding market demand. If the external demand is fewer than the internal capacity, it is known as an external constraint.

Throughput can be defined in the following formula:

$$\frac{\text{Sales Price} - \text{Variable Costs}}{\text{Time}}$$

Profits should be understood when dealing with throughput.

Inventories are known as raw materials, unfinished goods, purchased parts, or any investments made. Inventory should be seen as dollars on shelves. Any inventory is a waste unless utilized in a just in time manner.

Operating expenses should include all expenses utilized to produce a good. The less the operating expenses, the better. These costs should include direct labor, utilities, supplies, and depreciation of assets.

Applying the TOC concept helps guide making the weakest link stronger. There are five steps to the process of TOC:

1. Identify the constraint or the weakest link.
2. Exploit the constraint by making it as efficient as possible without spending money on the constraint or considering upgrades.
3. Subordinate everything else to the constraint – adjust the rest of the system so the constraint operates at its maximum productivity. Evaluate the improvements to ensure the constraint has been addressed properly and it is no longer the constraint. If it is still the constraint, complete the steps, otherwise skip Step 4.
4. Elevate the constraint – This step is only required if step two and three were not successful. The organization should take any action on the constraint to eliminate the problem. This is the process where money should be spent on the constraint or upgrades should be investigated.
5. Identify the next constraint and begin the five-step process over. The constraint should be monitored and continuous improvement should be completed.

Single Minute Exchange of Dies

What is SMED?

Single Minute Exchange of Dies (SMED) consists of the following:

- Theory and set of techniques to make it possible to perform equipment setup and changeover operations in under ten minutes
- Originally developed to improve die press and machine tool setups, but principles apply to changeovers in all processes
- It may not be possible to reach the "single-minute" range for all setups, but SMED dramatically reduces setup times in almost EVERY case
- Leads to benefits for the company by giving customers variety of products in just the quantities they need
- High quality, good price, speedy delivery, less waste, cost effective

It is important to understand large lot production which leads to trouble.

The three key topics to consider when understanding large lot production are the following:

- Inventory waste
  - Storing what is not sold costs money
  - Ties up company resources with no value to the product
- Delay
  - Customers have to wait for the company to produce entire lots rather than just what they want
- Declining quality
  - Storing unsold inventory increases chances of product being scrapped or re-worked, adding costs

Once this is realized, the benefits of SMED can be understood:

- Flexibility
  - Meet changing customer needs without excess inventory
- Quicker delivery
  - Small-lot production equals less lead time and less customer waiting time
- Better quality
  - Less inventory storage equals fewer storage-related defects
  - Reduction of setup errors and elimination of trial runs for new products

- Higher productivity
    - Reduction in downtime
    - Higher equipment productivity rate

Two types of operations are realized during setup operations which consist of internal and external operations. Internal setup is a setup that can only be done when the machine is shut down (e.g., a new die can only be attached to a press when the press is stopped).

External setup is a setup that can be done while the machine is still running (e.g., bolts attached to a die can be assembled and sorted while the press is operating).

It is important to convert as much internal work as possible to external work, as shown in Figure 8.64.

Four important questions to ask yourself when understanding SMED are the following:

- How might SMED benefit your factory?
- Can you see SMED benefiting you?
- What operations are internal operations?
- What operations are external operations?

There are three stages to SMED which are defined below:

- Separate internal and external setup
    - Distinguish internal vs external
    - By preparing and transporting while the machine is running can cut changeover times by as much as 50%
- Convert internal setup to external setup
    - Re-examine operations to see whether any steps are wrongly assumed as internal steps
    - Find ways to convert these steps to external setups

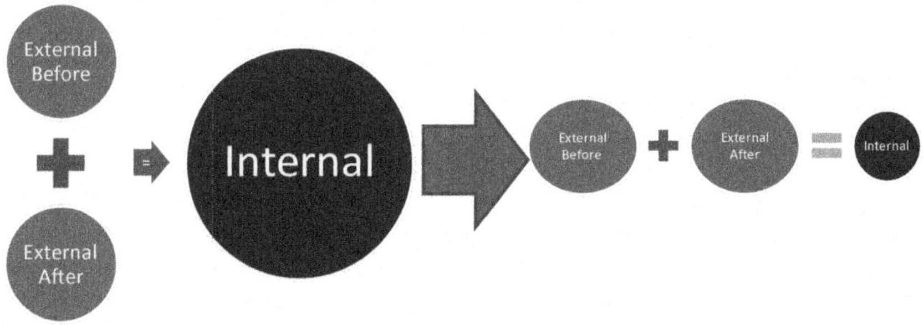

FIGURE 8.64
Internal versus external setup.

- Streamline all aspects of setup operations
    - Analyze steps in detail
    - Use specific principles to shorten time needed especially for steps internally with machine stopped

Five traditional setup steps are also defined:

- *Preparation* – Ensures that all the tools are working properly and are in the right location.
- *Mounting and Extraction* – Involves the removal of the tooling after the production lot is completed and the placement of the new tooling before the next production lot.
- *Establishing Control Settings* – Setting all the process control settings prior to the production run. Inclusive of calibrations and measurements needed to make the machine, tooling operate effectively.
- *First Run Capability* – This includes the necessary adjustments (re-calibrations, additional measurements) required after the first trial pieces are produced.
- *Setup Improvement* – The time after processing during which the tooling, machinery is cleaned, identified, and tested for functionality prior to storage.

In order to determine the proportion of current setup times, the following chart shown in Table 8.8 can be completed.

The three stages of SMED are explained next.

TABLE 8.8

Proportion of Setup Times before SMED Improvements

Setup Steps	Setup Type Traditional Internal	Setup Type Traditional External	Resource Consumption (%)	Setup Type One Step Internal	Setup Type One Step External
Preparation	x		20		x
Mounting and extraction	x		5	x	
Establish control settings	x		15		x
First run capability	x		50	N/A	N/A
Process improvement	x		10		x

Description of Stage 1 – Separate Internal vs External Setup

Three techniques help us separate internal vs external setup tasks:

1. Use checklists
2. Perform function checks
3. Improve transport of die and other parts

Checklists: A checklist lists everything required to set up and run the next operation. The list includes items such as:

- Tools, specifications, and workers required
- Proper values for operating conditions such as temperature, pressure etc.
- Correct measurement and dimensions required for each operation
- Checking item of the list before the machine is stopped helps prevent mistakes that come up after internal setup begun

An operation checklist is shown in Table 8.9.

Function Checks

- Should be performed before setup begins so that repair can be made if something does not work right.
- If broken dies, molds, or jigs are not discovered until test runs are done, a delay will occur in internal setup.
- Make sure such items are in working order before they are mounted will cut down setup time a great deal.

Improved Transport of Parts and Tools

- Dies, tools, jigs, gauges, and other items needed for an operation must be moved between storage areas and machines, then back to storage once a lot is finished.

- To shorten the time the machine is shut down, transport of these items should be done during external setup.
- In other words, new parts and tools should be transported to the machine before the machine is shut down for changeover.

TABLE 8.9

Operation Checklist

Operation Checklist			Effective: Nov-11
Equipment:			
Operation:			
Date:			
	Employees Trained for Setup and Operations (Need Two People)		
	Name of Employee	x	Name of Employee
x	Name of Employee		Name of Employee
	Tools Needed		
	Automatic Nut Driver		
x	Hex Wrench		
x	Rolling Cart		
	Tool		
	Tool		
	Tool		
	Tool		
x	Tool		
	Parts Needed		
x	Elevator Plate – 3.5 lb. Size		
x	Compression Plate – 3.5 lb. Size		
x	Feed Auger		
	Part		
	Part		
	Part		
	Part		
	Standard Operating Procedure to Follow		
x	SOP 001 – Changeover Procedure		
x	SOP 003 – Cleandown		
	Procedure		
	Procedure		
	Procedure		

Description of Stage 2 – Convert Internal Setups to External Setups

I-Advance preparation of conditions

- Get necessary parts tools and conditions ready before internal setup begins.
- Conditions like temperature, pressure, or position of material can be prepared externally while the machine is running (i.e., pre-heating of mold/material).

II-Function standardization:

- It would be expensive and wasteful to make external dimensions of every die, tool or part the same, regardless of the size or shape of the product it forms. Function standardizations avoid this waste by focusing on standardizing only those elements whose functions are essential to the setup.
- Function standardization might apply to dimensioning, centering, securing, expelling, or gripping.

III-Implementing Function standardization with two steps:

- Look closely at each individual function in your setup process and decide which functions, if any, can be standardized.
- Look again at the functions and think about which can be made more efficient by replacing the fewest possible parts (e.g., clamping function standardization).

Internal vs external setups can be put as provided in Table 8.10.

Description of Stage 3 – Streamline ALL Aspects of the Setup Operation

- External setup improvement includes streamlining the storage and transport of parts and tools.
- In dealing with small tools, dies, jigs, and gauges, it is vital to address issue of tool and die management.

TABLE 8.10

Internal versus External Setup Table

Internal vs External Setups			
Classify Items under Each Category			
	Internal		**External**
1		1	
2		2	
3		3	
4		4	
5		5	
6		6	
7		7	
8		8	
9		9	
10		10	
Which items would you convert from internal to external setup?			
1		1	
2		2	
3		3	
4		4	
5		5	
Why?			
1			
2			
3			
4			
5			

Ask Questions Like

- What is the best way to organize these items?
- How can we keep these items maintained in perfect condition and ready for the next operation?
- How many of these items should we keep in stock?

Improving Storage and Transport

- Operation for storing and transporting dies can be very time consuming, especially when your factory keeps a large number of dies on hand.
- Storage and transport can be improved by marking the dies with color codes and location numbers of the shelves where they are stored.

Streamlining Internal Setup

- Implement parallel operations, using functional clamps, eliminating adjustments, and mechanization

Implementing Parallel Operations

Machines such as plastic molding machines and die casting machines often require operation at both the front and back of the machine. One person changeovers of such machines mean wasted time and movement because the same person is constantly walking back and forth from one end of the machine to the other.

Parallel operations divide the setup operation between two people, one at each end of the machine. When setup is done using parallel operations, it is important to maintain reliable and safe operations and minimize waiting time. To help streamline parallel operations, workers should develop and follow procedural charts for each setup

A Setup Conversion Matrix is shown in Table 8.11.

The final understanding of SMED comes from basic principles such as observing with videos.

If there is nobody in the screen, it means there is waste present.

It is important to understand that SMED is more than just a series of techniques. It is a fundamental approach to improvement activities. A personal action plan should be found to adhere to each business' needs. It important to find ways to implement SMED into environments to continue the sustainability of the businesses. To begin the process, a communication plan should be implemented.

TABLE 8.11

Setup Conversion Matrix

Setup Conversion Matrix							
Sheet					Date:	Page	of
Area/Department	Machine/Equipment Name	Setup Tools Required	Operator Number		Standard Setup Time		
			Date Prepared		Minutes		
CURRENT PROCESS		**CURRENT TIME**		**IMPROVEMENT**	**PROPOSED TIME**		
NO.	Task/Operation	Internal	External		Internal	External	
	Current Total			**Improve Total**			

Conversation Methodology

Preparation of Setup Process — Combining Equipment Functionality — Standardized Jigs

Total Productive Maintenance

Total productive maintenance (TPM) has been a well-known activity that has several names associated within. Many people associate TPM with total predictive maintenance or total preventative maintenance. The association explained below will be total productive maintenance, but includes the total preventive maintenance as well.

TPM is performed in the improve phase based on downtimes or efficiency losses. The downtimes associated can be planned or unplanned. The goal of TPM is to increase all operational equipment efficiencies to above 85% by eliminating any wasted time such as setup time (see SMED section), idle times, downtimes, start-up delays, and any quality losses.

TPM ensures minimal downtime, but in turn requires no defects as well. There are three basic steps for TPM that have several steps within each.

1. Analyze the current processes
   a. Calculate any costs associated with maintenance
   b. Calculate Overall Equipment Effectiveness (OEE) by finding the proportion of quality products produced at a given line speed
2. Restore equipment to its original and high operating states
   a. Inspect the machinery
   b. Clean the machinery
   c. Identify necessary repairs on the machinery
   d. Document defects
   e. Create a scheduling mechanism for maintenance
   f. Ensure maintenance has repaired machinery and improvements are sustained
3. Preventative maintenance to be carried out
   a. Create a schedule for maintenance with priorities – include high machinery defects, replacement parts, and any information pertaining
   b. Create stable operations – complete root cause analysis on high machinery defects and machinery that causes major downtime
   c. Create a planning and communication system – documentation of preventative maintenance activities should be accessible to all people so planning and prioritization within is completed

d. Create processes for continuous maintenance – inspections should occur regularly and servicing for any machinery should be noted on a scheduled basis
e. Internal operations should be optimized – any internal operations should be benchmarked with improvements from other areas to eliminate time spent on root cause analysis. When defects of machinery are not understood, it is important to put the machinery back to its original state to understand the root causes more efficiently. Time to exchange parts or retrieve parts should also be minimized.
f. Continuous improvement on preventative maintenance – Train employees for early detection of problems and maintenance measures. Visual controls should be put in place for changeovers. 5S should take place to eliminate wasted time. The documentation should be communicated and plans should be given regularly. All aspects should be looked upon to see if continuous improvements can be made.

The key TPM indicators will be able to show the following main issues:

- OEE
- Mean time between failures
- Mean time to repair

TPM is crucial to sustainability because it involves all the employees including high level managers and creates planning for preventative maintenance so the issues are fixed before they become an error or defect. TPM also is a journey for educating and training the work force to be familiar of machinery, parts, processes, and damages while being productive.

Design for Six Sigma

Design for Six Sigma (DFSS) is another methodology that includes five phases called Define Measure Analyze Design Verify (DMADV) which stands for the following:

The difference of DMADV from DMAIC is the design and verification portions. DMAIC is process improvement driven whereas DMADV is for designing new products or services. Design stands for the designing of new processes required including the implementation.

Verify stands for the results being verified and the performance of the design to be maintained.

The purpose of DFSS is very similar to the regular DMAIC cycle where it is a customer driven design of processes with Six Sigma capabilities. DFSS does not only have to be manufacturing driven, the same methodologies can be used in service industries. The process is top-down with flow down CTQs that match flow up capabilities. DFSS is quality based where predictions are made regarding first pass quality. The quality measurements are driven through predictability in the early design phases. Process capabilities are utilized to make final design decisions.

Finally, process variances are monitored to verify Six Sigma customer requirements are met.

The main tools utilized in DFSS are FMEAs, quality function deployment (QFD), design of experiments (DOE), and simulations.

Quality Function Deployment

Dr. Yoji Akao developed quality function deployment (QFD) in 1966 in Japan. There was a combination of quality assurance and quality control that led to value engineering analyses. The methods for QFD are simply to utilize consumer demands into designing quality functions and methods to achieve quality into subsystems and specific elements of processes. The basis for QFD is to take customer requirements from the Voice of the Customer and relay them into engineering terms to develop products or services. Graphs and matrices are utilized for QFD. A house type matrix is compiled to ensure the customer needs are being met in the transformation of the processes or services designed (Figure 8.65).

The QFD house is a simple matrix where the legend is used to understand quality characteristics, customer requirements, and completion.

Design of Experiments

Design of experiments (DOE) is an experimental design that shows what is useful, what is a negative connotation, and what has no effect. The majority of the time, 50% of the factors have no effect.

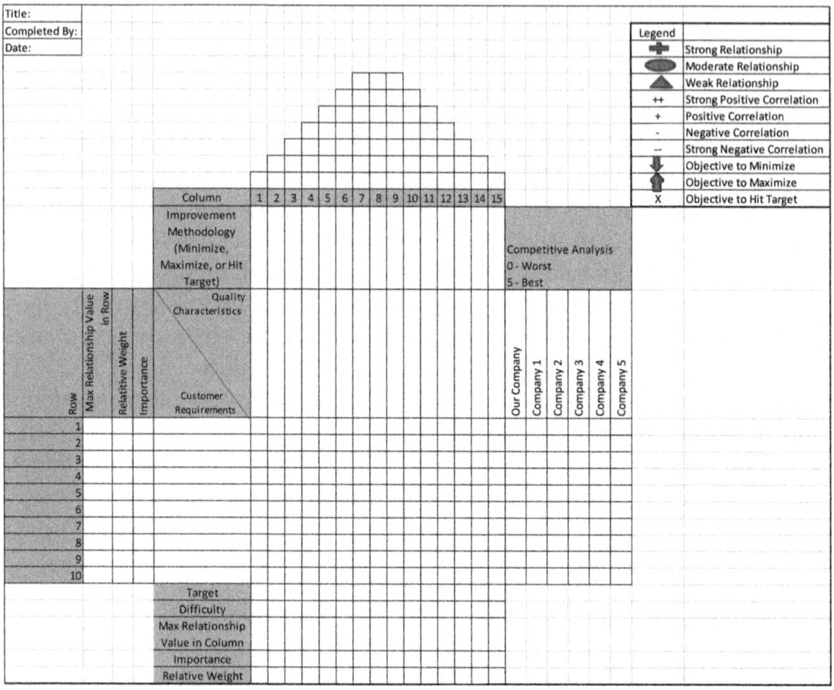

FIGURE 8.65
Quality function deployment house.

DOEs require a collection of data measurements, systematic manipulation of variables also known as factors placed in a pre-arranged way (experimental designs), and control for all other variables. The basis behind DOEs is to test everything in a pre-arranged combination and measure the effects of each of the interactions.

The following DOE terms are used:

- Factor: An independent variable that may affect a response.
- Block: A factor used to account for variables that the experimenter wishes to avoid or separate out during analysis.
- Treatment: Factor levels to be examined during experimentation.
- Levels: Given treatment or setting for an input factor.
- Response: The result of a single run of an experiment at a given setting (or given combination of settings when more than one factor is involved).

- Replication (Replicate): Repeated run(s) of an experiment at a given setting (or given combination of settings when more than one factor is involved).

There are two types of DOEs: full-factorial design and fractional factorial design.

Full-factorial DOEs determine the effect of the main factors and factor interactions by testing every factorial combination.

A full-factorial DOE tests all factors and their levels combined with one another covering all interactions. The basic design of a three-factorial DOE is shown in Table 8.12.

The effects from the full-factorial DOE can then be calculated and sorted into main effects and effects generated by interactions.

$$\begin{aligned}\text{Effect} = {} & \text{Mean Value of Response when Factor Setting} \\ & \text{is at High Level } (Y_A +) \\ & - \text{Mean Value of Response when Factor Setting} \\ & \text{is at Low Level } (Y_A -)\end{aligned}$$

In a full-factorial experiment, all of the possible combinations of factors and levels are created and tested.

In a two-level design (where each factor has two levels) with k factors, there are 2^k possible scenarios or treatments.

- Two factors each with two levels, we have $2^2 = 4$ treatments.
- Three factors each with two levels, we have $2^3 = 8$ treatments.
- k factors each with two levels, we have 2^k treatments.

TABLE 8.12

Full-Factorial DOE

	Factors			Factor Interactions			
Number	**A**	**B**	**C**	**AB**	**AC**	**BC**	**ABC**
1	–	–	–	+	+	+	–
2		–	–	–	–	+	+
3	–	+	–	–	+	–	+
4	+	+	–	+	–	–	–
5	–	–	+	+	–	–	+
6	+	–	+	–	+	–	–
7	–	+	+	–	–	+	–
8	+	+	+	+	+	+	+

The analysis behind the DOE consists of the following steps:

1. Analyze the data
2. Determine factors and interactions
3. Remove statistically insignificant effects from the model such as p-values of less than .1 and repeat the process
4. Analyze residuals to ensure the model is set correctly
5. Analyze the significant interactions and main effects on graphs while setting up a mathematical model
6. Translate the model into common solutions and make sustainable improvements

A fractional factorial design locates the relationship between influencing factors in a process and any resulting processes while minimizing the number of experiments. Fractional factorial DOEs reduce the number of experiments while still ensuring the information lost is as minimal as possible. These types of DOEs are used to minimize time spent, money spent, and eliminating factors that seem unimportant.

The formula for a fractional factorial DOE is:

2^{k-q}, where q equals the reduction factor.

The fractional factorial DOE requires the same number of positive and negative signs as a full-factorial DOE.

The fractional factorial DOE is shown in a matrix in Table 8.13.

Mood Median Test

Mood's Median test compares the medians of different samples of data when non-normal data is present and there are obvious outliers in the data.

Example 1 Mood's Median Test

```
Mood Median Test: Temperature versus Location
Mood median test for Temperature
Chi-Square = 17.07      DF = 1      P = 0.000
                                Individual 95.0% CIs
Location N<= N> Median Q3-Q1 --+---------+---------+---------+----
San Francisco 7  23    1.1282  0.0306             (----*----)
Cleveland    23   7   1.0882   0.0482   (-----*-----)
                                --+---------+---------+---------+----
                               1.080      1.100     1.120     1.140
```

TABLE 8.13

Fractional Factorial DOE

	Fractional Factorial Matrix											D				
Run	**(I)**	**A**	**B**	**C**	**D**	**AB**	**AC**	**AD**	**BC**	**BD**	**CD**	**ABC**	**ABD**	**ACD**	**BCD**	**ABCD**
1	+	−	−	−	−	+	+	+	+	+	+	−	−	−	−	+
2	+	+	−	−	+	−	−	+	+	−	−	+	−	−	+	+
3	+	−	+	−	+	−	+	−	−	+	−	+	−	+	−	+
4	+	+	+	−	−	+	−	−	−	−	+	−	−	+	+	+
5	+	−	−	+	+	+	−	−	−	−	+	+	+	−	−	+
6	+	+	−	+	−	−	+	−	−	+	−	−	+	−	+	+
7	+	−	+	+	−	−	−	+	+	−	−	−	+	+	−	+
8	+	+	+	+	+	+	+	+	+	+	+	+	+	+	+	+

```
Overall median = 1.1127

A 95.0% CI for median(San Francisco) - median(Cleveland):
(0.0277,0.0523)
```

Based on the data in the subgroups, the difference between the subgroup medians is 0.04 (1.1282–1.0882). The confidence interval for the difference provides more detail and confirms (with 95% confidence), that the difference in medians is somewhere between 0.0277 and 0.0523. The *p*-value for the test is 0.0. Since this is less than an alpha level of 0.05 we can say, with 95% confidence, that the medians of the subgroups are different. A rough graph of the 95% confidence intervals (CIs) for the medians of the subgroups is shown. The statistical conclusion is that the medians are different.

To summarize the results, we can be very confident that there is a difference in the median ratios for temperature between San Francisco and Cleveland.

Example 2 Mood's Median Test

```
Mood Median Test: Water Solubility versus Location
Mood median test for Water Solubility
Chi-Square = 0.07DF = 1P = 0.796
                                Individual 95.0% CIs
Location N<=N> MedianQ3-Q1------+---------+---------+---------+
San Francisco 15 15 5.73   1.84     (-*-----------------------)
Cleveland     16 14 5.70 2.04(*-------------------------------)
                           ------+---------+---------+---------+
                                6.00      6.60      7.20      7.80

Overall median = 5.70

A 95.0% CI for median(San Francisco) - median(Cleveland):
(-1.82,1.43)
```

Based on the data in the subgroups, the difference between the subgroup medians is 0.03 (5.73–5.7). The confidence interval for the difference provides more detail and confirms (with 95% confidence), that the difference in medians is somewhere between –1.82 and 1.43. The *p*-value for the test is 0.796. Since this is more than an alpha level of 0.05 we can say, with 95% confidence, that the medians of the subgroups are not different. A rough graph of the 95% confidence intervals (CIs) for the medians of the subgroups is shown above with overlap. The statistical conclusion is that the medians are the same.

To summarize the results, we can be very confident that there is no difference in the median ratios for water solubility between San Francisco and Cleveland.

Control Plans

A control plan is a vital part of sustainability because without it, there is no sustainability. A control plan takes the improvements made and ensures that they are being maintained and continuous improvement is achieved. A control plan is a very detailed document that includes who, what, where, when, and why (the why is based on the root cause analysis). The 12 basic steps of a control plan are listed below:

1. Collect existing documentation for the process
2. Determine the scope of the process for the current control plan
3. Form teams to update the control plan regularly
4. Replace short-term capability studies with long-term capability results
5. Complete control plan summaries
6. Identify missing or inadequate components or gaps
7. Review training, maintenance, and operational action plans
8. Assign tasks to team members
9. Verify compliance of actual procedures with documented procedures
10. Retrain operators
11. Collect sign-offs from all departments
12. Verify effectiveness with long-term capabilities

A control plan ensures consistency while eliminating as much variation from the system as possible. The plans are essential to operators because it enforces SOPs and eliminates changes in processes. It also ensures preventative maintenance (PM) is performed and the changes made to the processes are actually improving the problem that was found through the root cause analysis. Control plans hold people accountable if reviewed at least quarterly.

A sample control plan is shown in Figure 8.66.

Six Sigma Control Plan

Product:				Core Team:					Date (orig):		
Key Contact:									Date (revised):		
Phone:											
Process	Process Step	Input	Output	Process Specs (LSL, USL, Target)	Pkg./Date	Measurement Technique	%P/T	Sample Size	Sample Frequency	Control Method	Reaction Plan

FIGURE 8.66
Control plan.

Bibliography

Green Schools Initiative, The David Brower Center.

Leap, Local Energy Alliance Program, Charlottesville, VA (2010).

Water Use Efficiency Branch, Sacramento, CA (2010).

Commercial Buildings Initiative, Zero Energy Commercial Buildings Consortium (2008).

http://texasiof.ces.utexas.edu/PDF/Presentations/Nov4/CaseStudyFiorino.pdf

http://www.energy.ca.gov/process/pubs/toolkit.pdf

"The Story of Stuff", 2007: http://www.storyofstuff.com/

"Sustainability: Utilizing Lean Six Sigma Techniques" 2012, Tina Agustiady, Adedeji B. Badiru, CRC Press.

Index

Printed in the United States
by Baker & Taylor Publisher Services